ENAMELLED STREET SIGNS

by

Christopher Baglee & Andrew Morley

Everest House
Publishers   New York

Copyright © 1978 by Christopher Baglee and Andrew Morley.
All Rights Reserved.

Previously published in England as *Street Jewellery* by
New Cavendish Books, London.

This work or parts thereof may not be reproduced without prior
permission in writing from the publishers acting on behalf of the
copyright holders.

ISBN: 0 89696 055 2
Library of Congress Catalog Card Number: 78 74585.

Designed by Badge Group Design, Newcastle upon Tyne, in
        association with John Cooper.
Photography by R. Barker, Paul Davies, John Hawkins,
        Andrew Morley, Mike Ockenden, James Palm,
        Philipson Studios, Newcastle, and Peter Williams.

Historical and Technical Consultants, Vitreous Enamel
Development Council Ltd.

Printed and bound in England by Waterlow (Dunstable) Limited.
First American Edition.

Contents

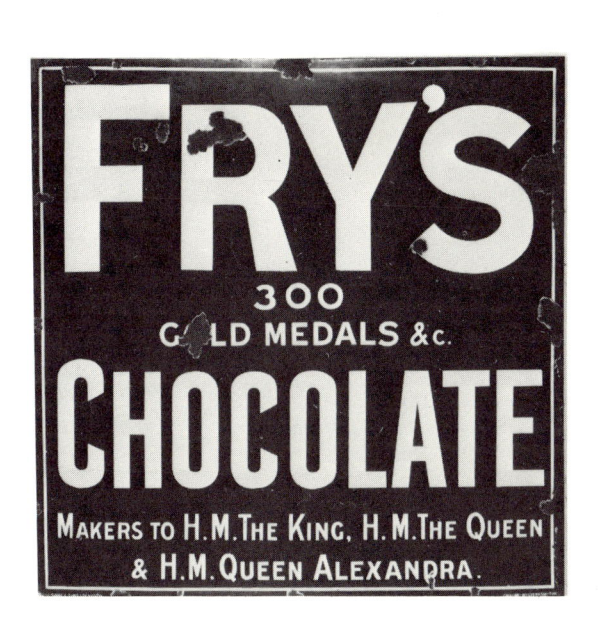

Preface, acknowledgements	6
Foreword by Geoffrey Clarke, F.C.I.S., Director, Vitreous Enamel Development Council Limited	8
Introduction	9
Historical background and manufacturing techniques	11
Locations and Social environment	19
Gallery of signs in colour	33
Design, Texture, Typestyles and Stencil Patterns	49
Collecting, Conservation, Restoration	59
Manufacturers past and present	69

Appendices:

I	Manufacturers	83
II	Bibliography	84
III	Museums and Public Displays	84
IV	Dealers	84
V	Advertising claims and slogans	85
Index – Text and single colour illustrations		86
Colour illustrations		87

As far as possible all illustrations within this book are to a scale of 1in : 1ft. Those illustrations bearing a white arrow do not conform to this scale.

Preface

Our interest in enamel signs developed independently in the early 1970s, when as professional artists and designers, we were attracted by the visual qualities of the few remaining signs extant in areas of Tyneside scheduled for demolition. An urge to preserve these artefacts, and a desire to bring together more of a similar nature, eventually led to our being rival collectors trying to obtain the same pieces. An offer from Richard Green, in mid 1976, to allocate space for an exhibition of enamel signs at the Laing Art Gallery, inspired us to pool our efforts and our collections, and to collaborate in this enterprise with renewed enthusiasm. We gathered as many examples as possible into our collection, but soon decided that we preferred to specialise solely in advertising enamels, although we recognised the potential of directional and informational enamel signs as a topic which has yet to be fully investigated. With the curiosity aroused by our activities, we set out to collate documentary information and photographic records, and soon felt that this material had the makings of a full-scale book, rather than a mere catalogue. This work is intended, therefore, not simply as an adjunct to the exhibition (now on national tour) but as the first survey of its kind on the subject, created for the general reader.

CB and AM
Newcastle on Tyne, 1978

Acknowledgements

Since commencing these two projects we have met with a refreshing response from members of the public, other collectors and from industry. We have also received the goodwill of hundreds of individuals and organisations, without whose help and encouragement we would have been unable to proceed. We owe a great deal to those authors and photographers who pioneered the publication of this subject through a number of magazine articles, and who have since unstintingly allowed us to draw from their researches.

In this respect we would like to thank, in particular, Geoffrey Clarke, Director of the Vitreous Enamel Development Council, and Ivor Beard for their valuable contributions to the Historical and Technical sections. Also we should like to give special thanks to Jim Palm, and Peter Williams for supplying many of their original photographs. Among all the others who helped in one way or another, the following have provided special assistance in the form of information, loans and gifts of signs, publicity, advice and the donation of sponsorship funds towards the exhibition (sometimes contributing in all these respects):

Brian Anderson, Badge Group Design, A.G. Barr (Tizer), Bass Museum, BBC Television, Beamish North of England Open Air Museum, Bluebell Railway, Blue Circle Group, BP Oil, Brooke Bond Oxo, David Brown, Bryant and May, Burmah-Castrol, Burnham Signs (Onyx), Causey Antiques, Nat Chaite, Colman Foods, Daily Express, Daily Mirror, Paul Davies, Dodo, Graham Douglas, John Dove, Enamelled Iron Signs, Esso Petroleum, Tom Falconer, Derek & Vera Faulkner, Liz Farrow, Finlay & Thompson, Stan Foster, Garnier Signs, Bob Gibson, Goodyear Tyre & Rubber, Erica Gorman, Richard Green, David Griffiths, Michael Harris, John Hawkins, Hawthorn Garage, Heinz, Hovis, Ingersoll, Irvine Motors, Ed Jaques, Philip John, Lens of Sutton, Stafford Linsley, Lyons, Mabel & Mr. Bennett, Bill McAlpine, Mills & Allen, Mobil Oil, Somerset Moore, City of Newcastle upon Tyne, N.E. Cooperative Society, Palethorpes, Past Times, Quaker Oats, Howard Raines, Reckitt & Colman, Sainsbury's, Scottish & Newcastle Breweries, Shell UK Oil, Adrian Shotton, Smiler, Spillers, Sterling Health, Jeff Stultiens, Frank Thornton, Tyne & Wear Industrial Monuments Trust, Tyne & Wear Museums Service, Tyne Tees Television, David Van der Plank, Victoria and Albert Museum, Vitreous Enamel Development Council, Volkswagen GB, Keith Wakenshaw, A.E. Warner, Martin White, Walter Willson, Wolverhampton Central Library, Wolverhampton Polytechnic Department of Visual Communications and David Wood.

Thanks are also due to all our friends and relations (particularly Mavis Baglee) for their unfailing sympathy, patience, interest and support during the frequently hectic and fraught preparation of the book and the exhibition. Thanks should also be given to the galleries around Britain for allowing us to display our collection to the public and to New Cavendish Books for their complete cooperation and encouragement.

Foreword

By Geoffrey Clarke, F.C.I.S.,
Director, Vitreous Enamel Development Council Limited

In these days we fall easily into a nostalgic mood. It seems we enjoy thinking about the past more than we do about the present or future. Take television, for example, that mirror of public taste. How many serialised plays and documentary features, and indeed, television advertisements, are about bygone days?

Nothing recalls the so-called good old days quite so vividly as enamelled metal signs, true relics of the past. Here, before our eyes, are steam tractors and threshing machines, horse-drawn pantechnicons and the spindly first generation of motor cycles. Those were the days when there was a booming market for tonic wines for the weakly, embrocations for horses and grate polish for solid iron stoves, when carpets were beaten instead of vacuum-cleaned and clothes mangled rather than spun. Nice to think about in this push-button age and to marvel at advertisements for daily newspapers at one old penny and twelve pint bottles of stout for half-a-crown — about the price of half-a-pint of bitter today.

There are now a number of private collections of these old enamel signs and some have even been received by public museums. Other colourful signs still proudly proclaim their message on private railways, such as the Bluebell Line in Sussex.

Longevity is not normally necessary in advertising placards today and, therefore, virtually everlasting enamel decoration is not much used in this field. However, enamelled signs are still being produced in fairly large quantities where durability and colour permanence is essential. The old enamel signs which have been saved from the scrap heap are really splendid advertisements — not least for their vitreous enamel finishes.

Christopher Baglee and Andrew Morley have brought together in this book a fund of information on old enamel advertising signs, and I am sure it will prove to be a most valuable addition to the general body of advertising literature.

Geoffrey Clarke

Introduction

Street jewellery, flashing in the winter sunlight, gleaming in gaslight, washed by the rain, impervious to the grime of industrial towns, a hard fact reinforcing its message through total permanence — 'The Plate That Outlasts All Others'; ruby red and emerald green, sapphire blue, ivory white and ebony black, an abundance of reassuring text and opulent imagery every time you visited the corner shop — the enamel sign.

Until their recent functional decline and the redevelopment of their sites, enamelled iron signs were, from the 1880s until the 1950s, amongst the most striking features of drab industrial streets in most towns and villages around the country. Most people do not notice the signs that remain, since they are now neglected, and their messages are often irrelevant. But from railway stations to post offices, back street corner shops to warehouses, enamel signs are still to be found. Clean the dirt from the surface and they will be as bright as the day when they were put up, probably some fifty to one hundred years ago. The advertisers' names still stir the memories of our grandparents' generation: 'We used Van Houten's Cocoa when I was a lad', and the claims such as 'Craven 'A' will not affect your throat' and 'Beecham's pills make all the difference' recall a time when the Trades Descriptions Act did not cramp the copywriter's style. We are surprised too by some of the price quotations such as 'Whitbread Ale and Stout — 2/6d a dozen' and 'Ingersoll watches from 5/-', which were peculiar to a pre-inflationary age as such signs would not be made in such a permanent medium today. The Patent Enamel Company, formed in 1880, claimed their enamelled sheet iron as 'The plate that outlasts all others'. In an age of iron, it was iron that served every function. The enamel sign was the logical extension of the Victorian preoccupation with permanence and stability. A sheet of steel, coated in coloured glass, fixed by huge iron staples to a solid brick or stone wall, was the ideal public relations exercise. These signs implied that not only the manufacturer and his product, but also his advertised claims, would last forever. The vogue for early advertising has brought in its wake a renewal of appreciation of the special qualities of the enamel sign.

This book attempts to present an outline of the historical, technical and visual elements of the enamel advertising medium.

10

Historical background and manufacturing techniques

HISTORICAL BACKGROUND

It was not until the early 1800s that porcelain enamelling on cast-iron was developed and practised in Central Europe. Later, around 1850, as sheet-iron manufacturing processes developed, porcelain enamelling was initiated in the United States and Britain.

In England, Benjamin Baugh, following a visit to Germany in 1857, started an enamelling business at premises in Bradford Street, Birmingham. Baugh took out a number of patents, from 1859 onwards, relating to the metal fabrication and enamelling process. He is described in the Patent Office register as 'Manager of Salt's Patent Enamel Works, Bradford Street, Birmingham'. It is understood that the first project undertaken by this company was making the decorated panels used on buildings and also on church altars. Decorated ceilings were supplied to the Gaekwar of Baroda, for the Durbar Hall in India, and also to the Kensington Museum and a London railway terminal. This firm is also known to have had a stand at the 1860 Exhibition in London.

Salt's became a public company and, in 1889 built a large factory at Selly Oak, which was designed and laid out for the manufacture of enamelled iron signs under the name of the 'Patent Enamel Company Limited'. This was probably the first and only factory built specifically for sign-making. It had twelve furnaces for fusing the enamel, two scaling furnaces as the iron required to be scaled and stretched, and a large printing room, plus a huge area for steampipe drying.
The company also smelted its own enamels and colour oxides. A railway siding ran into the factory, which also had its own canal arm and stables to accommodate horses.

A boom in enamelled signs came with the expansion of the railways, and orders for 100,000 signs were quite normal. Up to the time of World War I a vast quantity of signs were exported, but this market was lost as foreign countries, one by one, established their own enamelling firms.

Following Baugh's lead, other firms sprang up throughout Britain. Some of the largest, whose names can still be found in the small print on many signs were Chromo of Wolverhampton, Imperial of Birmingham, The Falkirk Iron Co., and Bruton's, Burnham's, Garnier's and Wood & Penfold, all of London. (See appendix 1)

It is recorded in Burnham's archives that when an advertiser commissioned a batch of signs, they usually specified that the sign-making company should be responsible for the distribution and fixing of the signs within a limited period from the time of manufacture. The railways facilitated distant distribution, which was then taken up by a local carrier with horse and cart, who would deliver and install the signs.

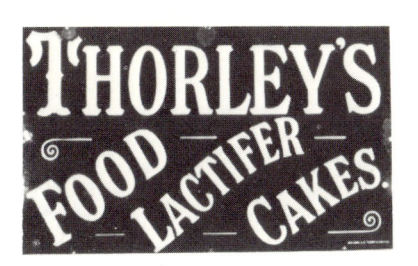

The work of distributing Colman's signs fell to their Tableting Staff, whose job it was to negotiate sites, fix the plates and clean and maintain them at regular intervals. Rents, either in money or in goods were paid to the property owner for the privilege. The Tableting Staff operated from three depots strategically placed throughout the country and these duties entailed full-time employment. Today the number of signs left *in situ* is very small. An article in *The Junior Traveller. Journey the Tenth*, printed and published by the Hovis Bread Flour Co. in the summer of 1902, deals with the different forms of Hovis advertising and promotion and reads: ' "Hovis food for babies and invalids" is a very nice plate 8 feet in length, with good clear letters that we supply for spaces under shop windows or any other bold position. We also have a similar plate for Hovis bread.'

Some of the earliest products to be advertised through the medium were animal foods — Thorley's, confectionary — Cadbury's, Fry's and Rowntree's, soaps — Hudson's, Pear's and tea-Mazawattee, Nectar and United Kingdom. Many high class stores and products were early users, e.g. Maples, Aquascutum and makers of pianos, but these prestigious signs seemed to disappear when the method become more widespread and vulgarised.

The hey-day of enamelled iron signs was comparatively short. They reached their peak before the 1914 War, went into decline from 1918, and their end was in sight by 1939. In the half century leading to World War II, millions of signs were produced. During the 1950s, however after a virtual halt of production in the previous decade, and with continued rationing of steel, there emerged a new and powerful rival in the form of huge hoardings, which not only masked the bomb sites, but also displayed the novel American-style photographic posters. World War II had struck the *coup de grace* to a dying industry by simultaneously destroying their main sites and thereby creating new show places for the rival medium. Furthermore, World War II and its aftermath, denied the sign manufacturers vital raw materials.

MANUFACTURING TECHNIQUES

The porcelain enamel process involves the re-fusing of powdered glass on a metal surface via two principal processes.

Cast-iron, dry process enamels, were the first to be used on a large scale. In this process the castings were sand blasted to give them a clean surface. The grip or ground coat was then applied, and this comprised a powdered glass, clay and water suspension, with a consistency similar to that of cream. This was dipped, slushed and sprayed onto the cool casting and allowed to dry.

The ware was then introduced into a furnace at about 900°C, and allowed to reach the temperature of the furnace. The hot ware was then withdrawn, and powdered glass dusted through a screen onto it. This powdered glass melted as it fell on to the surface of the ware and formed a continuous layer, known as 'enamel'. Several applications were made in this way, the ware being returned to the furnace for reheating before each application.

The enamelling of iron only became possible because of the ability to vitreous-enamel sheet wrought iron, and it was not until the 1920s, when Armco produced a steel sufficiently free from defects, that sheet steel could be used for signs. Since that time, all orders for steel to be vitreous-enamelled, have been for specially produced 'vitreous-enamelling quality steel'. Sheet iron was

Interiors of Burnham's factory, showing kiln (above) and screen printing beds (left).

ordered as 'pure scotch wrought puddled iron', and sign makers were very sad when the last 'hearth' of this was produced by the Pather Iron & Steel Co. Limited. There was less wastage with iron, since all defects could be rectified during processing and the colours on iron were crystal clear, whereas those on steel — when closely compared — seemed to have a greasy appearance. Incidentally, collectors can easily differentiate between iron and steel signs as those made from iron have backs of a mid-grey colour and will invariably have swill lines; the backs of steel signs are blue-black and normally by the time steel was used in production, the base colour was sprayed and not swilled. In the factory the 'grip' coat was never named as such; it was always referred to as 'grey' and not changed to blue-black.

Sheet steel enamelling has become the process most widely used for porcelain enamels nowadays. In this way the sheet is put through a cleaning and pickling process which prepares the surface and also receives a ground coat, containing a small percentage of cobalt to give adhesion. This ground coat is applied by the wet process, is dried and then fired in a furnace at about 830°C. After the ware has been removed from the furnace and cooled a second coat of cover enamel is applied. This cover coat may be of any desired colour and may have special requirements or properties, depending on what use the ware is to be put. It is commonly dipped or sprayed onto the ware and then fired at a slightly lower temperature. Enamel colours are achieved by adding metallic oxides to the ground 'frit' or glasses at some stage, in greater or lesser percentages, according to the strength of the colour desired. Essentially some will be 'softer' than others and will burn out at high — or additional — firing temperatures. Consequently, the ground coat will be hardened by being applied first and receiving the most firings, while the softest colour is applied last, to receive the minimum number of firings at the lowest temperature. Thus firing temperatures will vary from something over 860°C to as low as 800°C for one sign.

Finally, the decoration is applied and fired into the last coat. This is done with a rubber stamp coated with ceramic colours in the form of inks or by dusting on a coloured pigment to an area stamped with gum or varnish.

There is also a process of transfer which uses a ceramic decal which enables the image to be attached to and fired into the ware. These transfers are often needed where a small piece of fine, intricate work, such as a coat of arms, trademark or instructions, is essential on an otherwise wholly lettered sign. This technique was a comparative late comer and as lithographic processes could achieve similar results, was little used.

Originally the design was applied by the use of stencils, but today this is principally applied for one-off orders where the cost of making screens outweighs the more labour-intensive stencil process. In this stencil process the colour was sprayed on to the plate and after drying, had the consistency of weak distemper. The stencils, cut to the appropriate design, were placed on the plate and the exposed colour brushed away, leaving the design intact. The plate was then fired and the colour vitrified indelibly into the background. This process could be carried out with successive colours, using further stencils, until the most intricate designs and patterns were achieved. It was a process which demanded a great deal of skill, not only on the part of the stencil cutters, but also from those 'brushing out' as they had to work accurately and carefully ensuring that they only brushed away the material that was unwanted without disturbing the surface of the colour that was intended to remain.

A design too complicated for the stencilling process was etched on to a stone for each colour and the stone was then placed in a printing press. A print on paper was then taken and the sheet – still wet with ink – was laid over the sign and peeled off leaving the wet printing ink on the surface. Dry enamel pigment was then dusted over the ink, leaving the design in colour to be fired. This process was repeated for all the colours. In the twenties, instead of paper, gelatine sheets were used which could be cleaned each time and zinc plates were introduced to replace the stone.

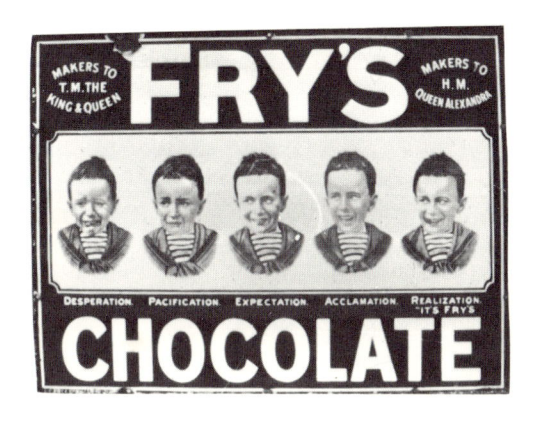

At the turn of the century a method of reproducing photographs in black or sepia on to vitreous enamel was invented. The monochrome Fry's Five Boys chocolate sign is an example of this process, while the coloured version is the same plate but tinted.

Today, however, much design application is carried out by the screen printing process. For this the enamel colour is specially formulated and ground into the form of ink which is then screened on to the steel plate. After screening the plate is fired. With this process it is possible to produce detail of great intricacy as well as multicoloured and pictorial designs, which would be impossible with the stencil method.

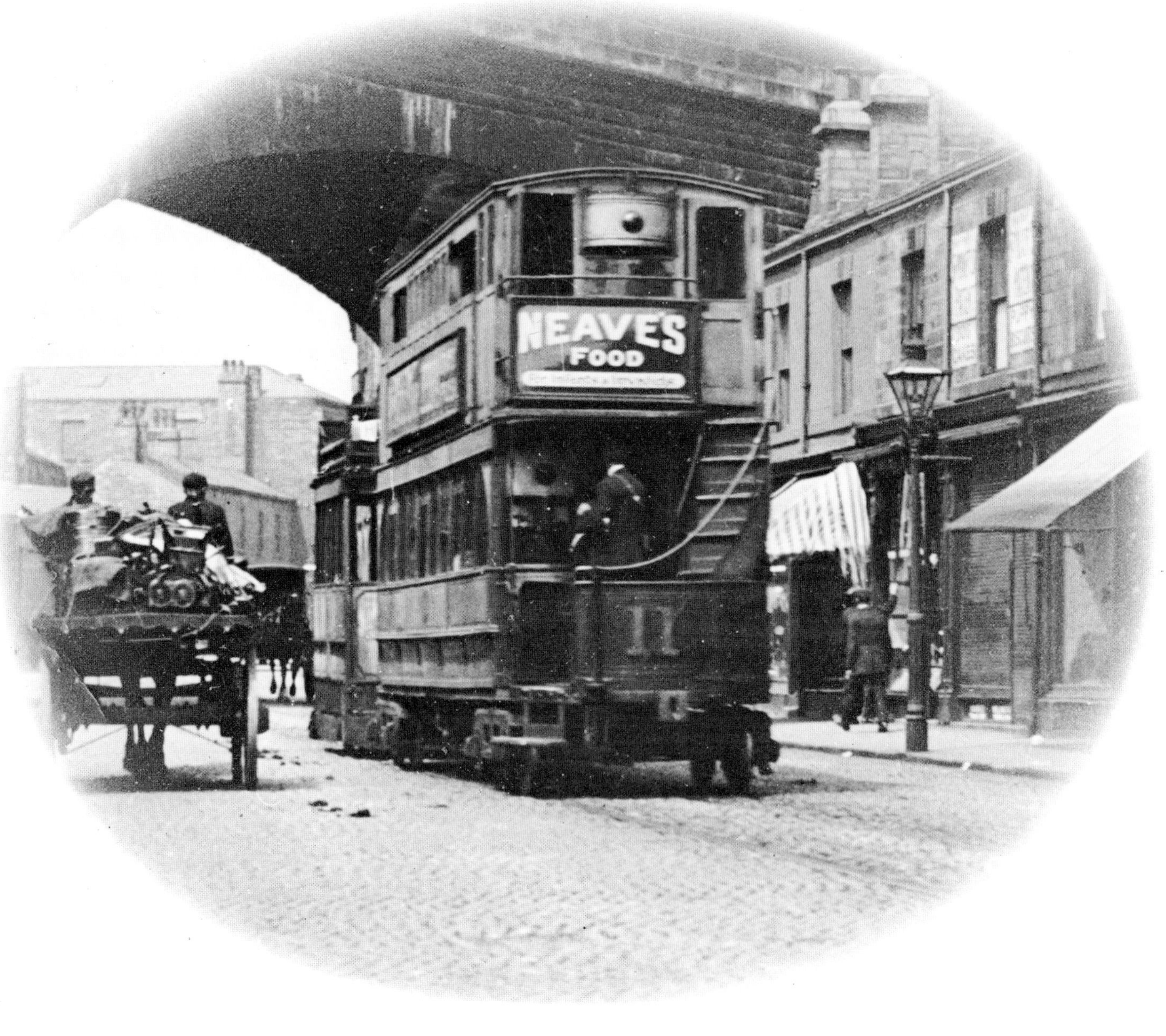
NEAVE'S
FOOD

Locations and Social environment

LOCATIONS

'The Daily Mirror — Best All Along The Line', reminds us of the wide range of locations in which these signs appeared. In this case, beside a railway track or on a station, it was a response to the expansion of the railways which contributed a great deal to the development of the sign making industry. The railways made possible the shipment of huge orders and provided ready made locations and a captive public. 'W.H. Lever was a great believer... he even went to the trouble of choosing the exact sites where the enamel signs were to be displayed at railway stations; and the advantage of the right-hand against the left-hand of the booking office was carefully considered' (noted by Robert Opie in *The Ephemerist*, Vol. 1, No. 10, May 1977, from Professor Wilson's *History of Unilever*). Virol saw the value of station approaches in the early days and acquired most of these, lining the entrance or exit routes with their standard blue, white and orange signs and adding such authoritative slogans as 'nursing mothers need it' and although it would be hard to find any other form of advertising used by Virol, everyone was aware of the product. This company jealously guarded their sites, even after World War II, until it was no longer viable for them to renew their leases. It was the change to higher rents for advertising sites on railway stations that lost these locations to the enamel sign and made them far more suitable for paper posters. The first company to recognise the value of stair risers as locations for signs was Mazawattee Tea, followed by Ovum and Redferns and they were imitated in turn, especially on trams and omnibuses, by Iron Jelloids.

Trams and omnibuses were the other main forms of transport favoured by advertisers as sign sites, but some commercial vehicles were also used, being clad liberally in custom-made signs. On the trams and buses the signs were fixed to the stair balustrading, the stair risers, the backs of seats and the outer bodywork panels. On the streets along which these vehicles travelled were located the major static placements. The motoring garages (usually converted smithies or stables) which valued the signs as evidence of their modernity, and most important — shop fronts. These were mainly corner shops, general dealers, post offices, ironmongers, pubs, off-licences and the gable ends, walls or fences of other prominent buildings, located in the natural commercial centres of cities, towns and villages. In the countryside signs for the specialist agricultural products such as Thorley's pig food, Toogood's seeds, and Ransome's farming equipment might be seen on barns and farmhouses, and often they still survive because of their isolation.

Today, the constantly changing images confronting us on the street billboards make the enamel sign an anachronism. It goes against the grain for an advertiser to retain the same campaign on a long term basis; to do so would indicate a lack of enterprise or a non-expanding market.

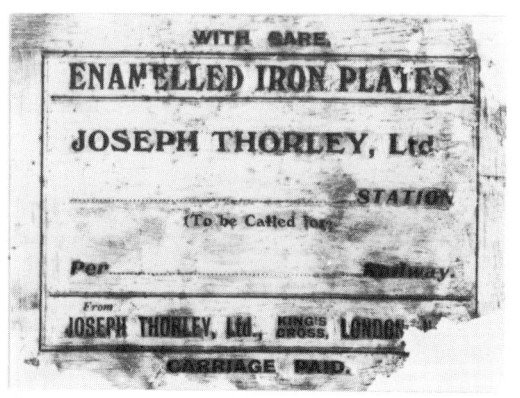

This advertising practice of bombarding public awareness with constantly changing imagery, exploded with a vengeance in Britain after World War II. At this time most major cigarette manufacturers deliberately removed their old advertising signs from the walls, as did Reckitt's with their Nugget Boot Polish signs — and many were replaced with a new dynamic phase of 'contemporary' Festival-of-Britain graphics. These were sometimes still on enamel, e.g. Turf cigarettes, but more usually they were on disposable and replaceable paper or tin. These were, of course, not unique to the 50s, having never declined in popularity as an advertising medium since the invention of printing, but with the development of silkscreen and photolithographic techniques, the paper poster came into its own again to fulfil demands more economically than enamel with its obdurate permanency in an age where obsolescence was structured into the manufacture and advertisement of so many products.

The Town and Country Planning Act of 1947 added a further restraint in that all signs fitted above fascia level required the complicated process of obtaining planning permission before they could be displayed. This was hardly profitable and so the alternative was for enamels to be relocated as trestle or forecourt signs.

The first L.G.O.C. Motor bus 1904.

21

LYONS' TEA
Buckley Furniture — Tel. 24706

BENSON and HEDGES
Pure gold...
FRY'S MILK CHOCOLATE

LYONS TEA
FILM OFFER
Player's Please

"TURF"
"TURF"
20 CIGARETTES 20
TRADE MARK
VIRGINIA CIGARETTES
"TURF"
VIRGINIA CIGARETTES
GUARANTEED MADE FROM TOBACCO AGED AT LEAST 3 YEARS IN THE WOOD

A.GODWARD
SMITH'S
HARDWARE
FRY'S
CHOCOLATE
FRY'S
CHOCOLATE

Reconstruction of a 1920's garage forecourt.

(National Motor Museum Photographic Library).

23

SIGNS ON THE MOVE
Pre-1914 public vehicle display

(Hulton Picture Library)

(London Transport Photographic Library)

24

(Courtesy London Transport)

(Courtesy London Transport)

25

(Courtesy London Transport)

(Courtesy London Transport)

26

(Courtesy London Transport)

(Courtesy London Transport)

27

(Courtesy London Transport)

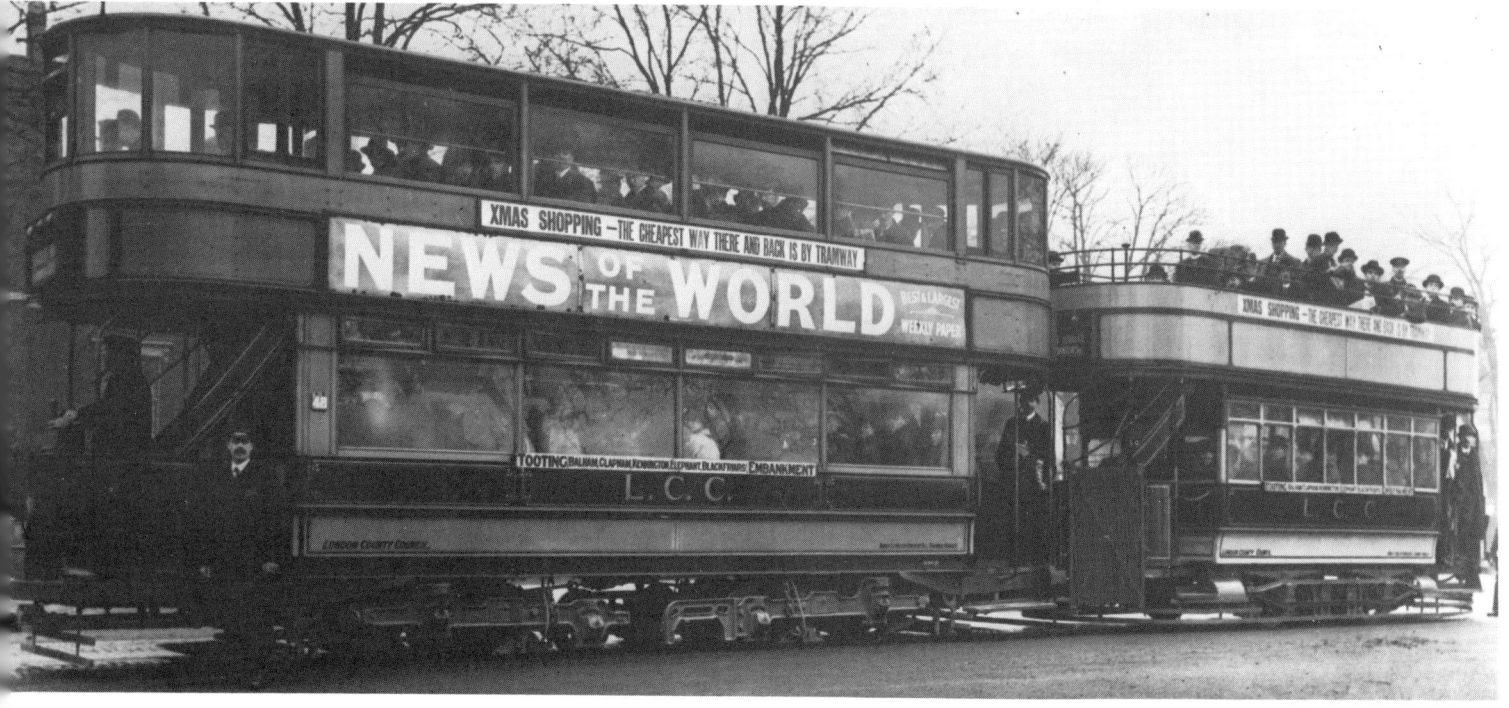

SOCIAL ENVIRONMENT

As evidence of changing social habits and conditions, the slogans and advertising claims on enamel signs are very instructive. A selection is given of some of the more exalted and outrageous of these, that could only have been made before the current age of advertising watchdogs.
'Craven 'A' will not affect your throat'
'Beecham's Pills make all the difference'
'Van Houten's Cocoa, best and goes farthest'*
'Epps's Cocoa for strong and weak'
'Fry's Cocoa – there is no better food'
Pears soap – 'Matchless for the complexion'
Oxo – 'Splendid with milk for children', 'Excellent with milk for growing children', 'Meat and drink to you', 'The perfect beef beverage'
Virol – 'Delicate children need it'
New Hudson cycles would climb any hill, but if you were not up to that kind of exertion you probably needed Wincarnis, which 'Gives New Life To The Invalid, New Strength To the Weak, Increased Vigour To Brain Workers, A Wealth of Health to Everyone. The World's Greatest Restorative In Cases of Anaemia: Depression: Brain Fag: Sleeplessness: Physical & Mental Prostration: Nerve Troubles: And in Convalescence.'

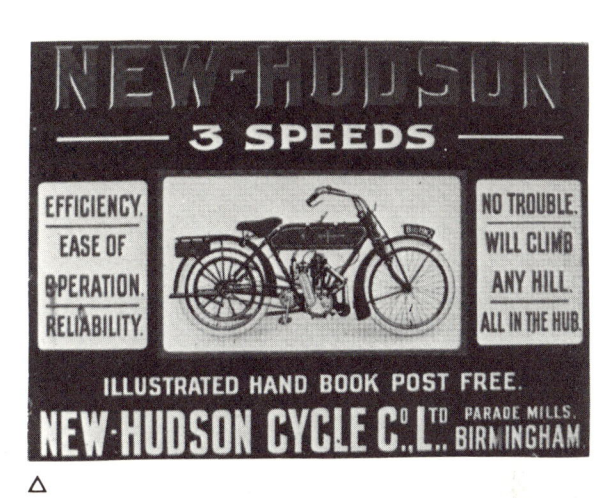

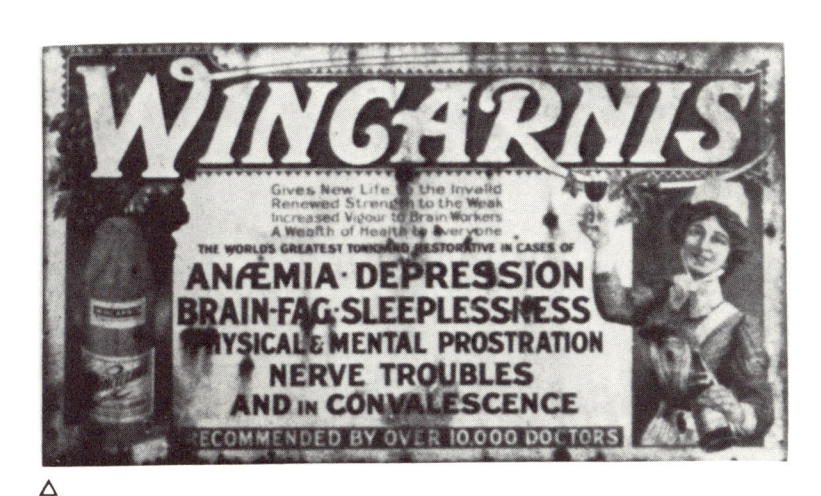

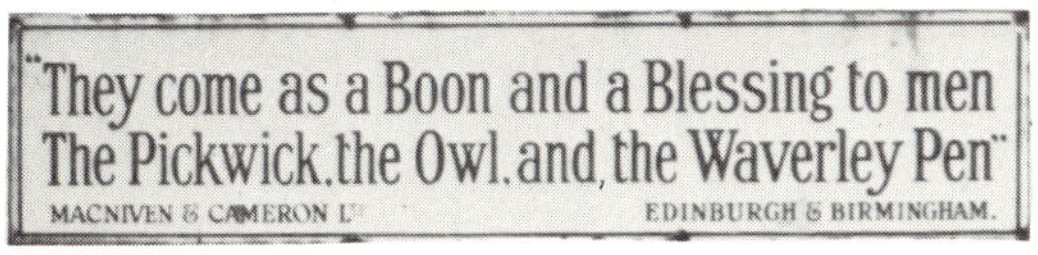

*James Blades, on page 53 of his autobiography, mentions a special use for this sign!

The next selection encouraged purchase by topical social comment, or appeal to class values:
'Bryant and May matches Support Home Industries, Employ British Labour'
'Simpson's whisky – as supplied to the House of Lords'
Hudson's soap 'For the People', 'Used in all "The Happy Homes Of England"'.
'Rinso, soak the clothes – that's all; saves coal every wash day'
'Nectar, the most economical tea sold'
'Lyon's tea, a packet for every pocket'
'Brooke Bond tea, spend wisely, save wisely'
'Lucas Batteries on extended credit terms.'

Thus the enamel sign reflected or created consumer demand both in colour and texture and unlike film or paper, remains an almost unblemished relic of its time.

The first area worth noting is where certain product types, frequently used and advertised at the turn of the century, have either been superseded by modern labour-saving products, or do not warrant present day advertising. Before the current wave of medical discovery and State care, there was a great demand for health-giving tonics such as Wincarnis and Anti-Laria (non-alcoholic sparkling wine, 'a stimulant second only to champagne, at a fraction of the cost'), ginger wine, Beecham's pills and Liver Salts of various kinds. Before the advent of electric washing machines and biological washing powders, there were signs advertising 'mangling done here' and many brands of soap ranging from Sunlight and Hudson's Dry Soap, to Mrs. Volvolutum's 'no rubbing, no soaking!'. The extensive use of other common household products recorded in enamel, but now reduced in importance are Colman's and Robin starch, Reckitt's blue, Chiver's carpet soap, several makes of metal polish, liquid Brasso, Komo metal paste and grate polishes such as Zebo, Zebra and James Dome Black lead. It was common to find advertisements for newly invented articles which are now very rarely advertised, such as Ediswan lamp bulbs or of obsolescent products such as Royal Daylight and Pratt's lamp oils and Veritas gas mantles, as well as several manufacturers of gas and oil engines for generating electricity and for farming purposes.

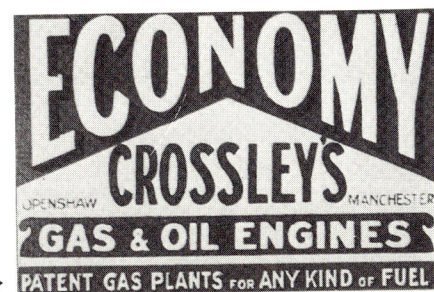

The emphasis on advertising various beverages has now changed completely. We no longer find cocoa greatly promoted, let alone as a valuable source of food. However, in the late nineteenth and early twentieth centuries, several rival companies, Cadbury's, Epps's, Fry's, Lyon's, Rowntree's and Van Houten's all brought their cocoa prominently to the notice of the public.

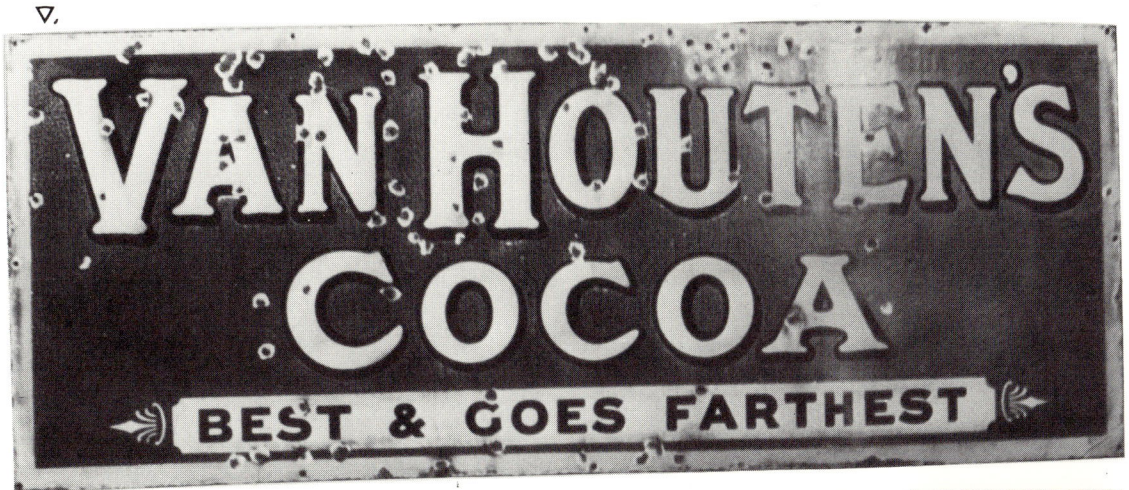

Nor is tea so widely publicised today, except by a few major companies, yet during the period mentioned previously, Blue Cross, Brooke Bond, Horniman's, Lipton's, Lyon's, Mazawattee, Melrose's, Nectar, Silverbrook, Typhoo, United Kingdom and many more were all in serious rivalry through the medium of the enamel sign. At the same time only one coffee company was blowing its trumpet — 'Don't be misled, Camp Coffee is the best.' — Competitors remained silent, hardly the case today!

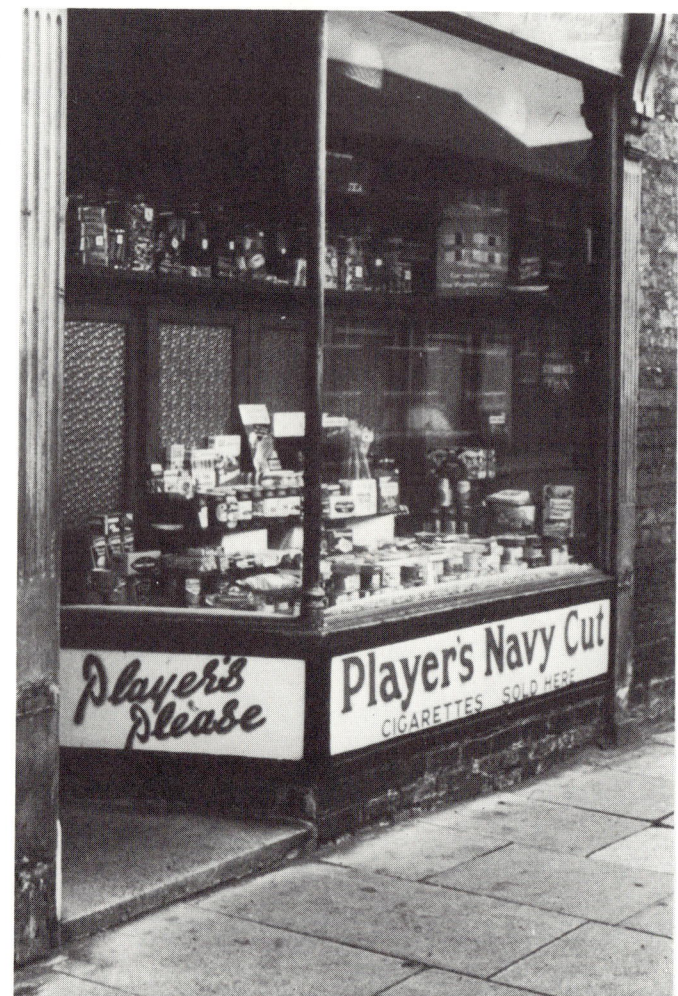

WILLS'S "CUT GOLDEN BAR" TOBACCO SOLD HERE
CADBURY'S COCOA ESSENCE
REGISTERED
Guaranteed Absolutely Pure Cocoa.
Absolutely pure
ST JULIEN
OGDEN'S ST JULIEN EMPIRE BLEND
EXQUISITE COOL AND FRAGRANT
TOBACCO
EMPIRE BLEND
Stephens
Ink
GREEN & LEDICOTT'S
MINERAL WATERS
Buy HUDSON'S Soap
HOME & FOREIGN REMOVAL EXPERTS
R.D.&J.B.FRASER LTD
FRASERS' HOUSE FURNISHERS IPSWICH
FRASERS' DEPOSITORIES & STRONG ROOMS IPSWICH
FRASERS IPSWICH
DEPOSITORIES FOR STORAGE OF FURNITURE
ESTIMATES FREE
IPSWICH
AND FELIXSTOWE
EAT COLMAN'S MUSTARD
WITH
OUR PRIME BEEF, MUTTON Etc.
AND ENJOY THEM MORE.
KARPOL
Buy it here
WAKEFIELD CASTROL MOTOR OIL
IT'S 'Meat & Drink' TO YOU
OXO CUBE
CONCENTRATED BEEF
Best of all -
Smoke WOODBINES
"MATCHLESS" METAL POLISH

Drink
TIZER
The Appetizer
MAKERS TO T.M. THE KING & QUEEN
FRY'S
MAKERS TO H.R.H. THE PRINCE OF WALES
300 GOLD MEDALS & DIPLOMAS
300 GOLD MEDALS & DIPLOMAS
DESPERATION. PACIFICATION. EXPECTATION. ACCLAMATION. REALIZATION. "IT'S FRY'S"
CHOCOLATE
Burgoyne's
Tintara, Ferruginous
BURGOYNE ON CORK
EX-LAX
chocolate laxative
keep 'regular' with
EX-LAX
PRESCRIPTIONS DRUGS TOILET ARTICLES
For the People
POWERFUL
HUDSON'S SOAP
EASY AND SAFE
Royal Daylight
ANGLO AMERICAN OIL Co. LTD
BEST AMERICAN LAMP OILS
LAMP OIL
MADE AT NEWCASTLE
SPILLERS
BALANCED
RATIONS
AND
FOR
SOAK THE CLOTHES-THAT'S ALL!
Rinso
SAVES COAL EVERY WASH-DAY
Inside Knowledge
THORLEY'S
FOOD
for PIGS
MACNIVEN & CAMERON'S PENS
3050 NEWSPAPERS
6d & 1/- per Box
DEPOT
LITTLE WOODS
BY APPOINTMENT
OXO
ONE CUBE MAKES A NOURISHING BEEF DRINK
DISSOLVE IN A CUP OF HOT WATER
OXO CUBE
ONE CUBE MAKES A NOURISHING BEEF DRINK
DISSOLVE IN A CUP OF HOT WATER
OXO CUBE
SPLENDID WITH MILK FOR CHILDREN

Wincarnis
The World's Greatest Wine Tonic and Nerve Restorative
Viking Milk
Viking Milk
Trade Mark
Latte Viking
Player's
Player's Navy Cut
Cigarettes "Medium"
Navy Cut Cigarettes
National Benzole Mixture
Wheel Brand
Gossages' Dry Soap
Sun Insurance Office
Drive with the Sun behind you
In Fine Powder
Hudson's Dry Soap
Used in all the "Happy Homes of England"
Hudson's
Brooke Bond dividend Tea
We sell sealed Shell
We sell
Lyons
2d Fruit Pies
Fry's
Makers to T.M. the King & Queen
Makers to H.R.H. the Prince of Wales
Pacification
Expectation
Acclamation
Realization "It's Fry's"
Chocolate
Palethorpes
Royal Cambridge

JOHN SINCLAIR'S
RUBICON
TWIST
Player's Please
"It's the Tobacco that Counts"
WESTWARD HO!
SMOKING MIXTURE
W.D.&H.O.WILLS.
ROWNTREE'S ELECT COCOA
MAKERS TO T.M. THE KING & QUEEN.
SPRATT'S
SPRATT'S 'builds up' a dog!
BONIO : MIXED OVALS : WEETMEET
BLUE BAND
MARGARINE
BRASSO
METAL
Polish.
MAKERS TO H.M.THE KING.
ROWNTREE'S
CHOCOLATES
AND
PASTILLES
MAKERS TO H.M.THE QUEEN.
WILLS'S
STAR
Cigarettes
W.D.&H.O.WILLS
Bristol & London
CIGARETTES
WILLS'S
WOODBINES
THOMSON'S
DYE
WORKS
PERTH
VERITAS MANTLES
DAGENITE
THE DEPENDABLE
ACCUMULATORS
SOLD HERE
Stephens Inks
FAHRENHEIT
CENTIGRADE
FOR ALL
TEMPERATURES
Stephens' Inks
CHURCHMAN'S
"TORTOISESHELL"
TORTOISESHELL
SMOKING MIXTURE
SMOKING MIXTURE.
FRESH
PALETHORPES
TO-DAY

CHIVERS'
CARPET SOAP 6d
BEFORE USE
AFTER USE
PER BALL
Sold
EVERYWHERE
FOR RESTORING THE COLOUR AND REMOVING GREASE OR INKSTAINS FROM CARPETS
BEST AND SIMPLEST CARPET CLEANER IN THE WORLD. MANUFACTORY BATH.
PRIZED IN ROYAL HOUSEHOLDS
Reckitt's
BLUE.
DEPÔT for NORFOLK
CHAMPION BOOTS
RELIABILITY GUARANTEED
HIS MASTER'S CHOICE
KENYA BEER
DEPOT FOR
"SWAN" INK
MORRIS
SERVICE
CARPETS CLEANED "SHAMPOOED" (SIMTUL PROCESS)
PATENT STEAM CARPET BEATING Co
CARPETS SHAMPOOED
PATENT STEAM CARPET BEATING Co
UP-TO-DATE EXPRESS DELIVERY BY OUR MOTOR VANS
PATENT STEAM CARPET BEATING Co Ltd
196, YORK ROAD, KING'S CROSS, N.
ROBIN STARCH
RALEIGH
THE ALL-STEEL BICYCLE
UNITED KINGDOM
TEA COMPANY'S
DELICIOUS TEAS
DIRECT FROM THE GROWERS
THE BEST TEAS THE WORLD PRODUCES
UNION CASTLE LINE
To SOUTH & EAST AFRICA
THE HIGHEST PINNACLE
For Quality and Flavour
PLAYER'S NAVY MIXTURE
MITCHELLS AND BUTLERS' ALES
(TRADE MARK)

"Speaks for itself"
SHARP'S
SUPER-KREEM
TOFFEE
Sir Kreemy Knut
SMOKE
OGDEN'S
"ROBIN"
10 CIGARETTES
CIGARETTES
Orders
Received for
P.&R.HAY
Established Over 100 Years.
Dyers & French
Cleaners
Leith Walk,
Edinburgh.
EDISWAN
THE
KING
OF
LAMPS
ROYAL
"EDISWAN"
From all Electricians, Ironmongers & Stores.
LAMBERT & BUTLER'S
RHODIAN
CURLY CUT
RHODIAN
CURLY CUT
a good all round
Tobacco
GARGOYLE
Mobiloil
"BB"
BRITISH MATCHES FOR BRITISH HOMES
PUCK
MATCHES
BRYANT & MAY LTD
BOVRIL
Makes
Contented
Cooks
Stephens'
Inks
ST BRUNO
OGDEN'S
ST BRUNO
FLAKE
EMPIRE BLEND
ROUGH CUT
FLAKE
EMPIRE BLEND
HUDSON'S
SOAP
SMOKE
PLAYER'S
NAVY CUT
TOBACCO
AND
CIGARETTES
Drink
TIZER
The Appetizer

SPILLERS SHAPES
for all dogs

Sunlight soap

QUAKER OATS
PURE

For your throat's sake smoke
CRAVEN "A" VIRGINIA CIGARETTES
They never vary!

WHS

SILVER SHRED
ROBERTSON'S
Lemon Marmalade

MOTOR **"BP"** SPIRIT

LUCAS BATTERIES ON EXTENDED CREDIT TERMS

Here take Health
Bermaline Bread
GRATIFIES & SATISFIES

OGDEN'S ST BRUNO THE STANDARD DARK **FLAKE**

DAY & MARTIN'S BLACK OR BROWN IN TINS & BOTTLES **POLISHES**

LYONS' TEA SOLD HERE

VENOS Lightning **COUGH CURE** for COUGHS, COLDS, FLU, BRONCHITIS, ASTHMA, & Children's Coughs

FRY'S CELEBRATED **CHOCOLATE**
MAKERS TO T.M. THE KING & QUEEN
MAKERS TO H.M. QUEEN ALEXANDRA

GOOD·YEAR

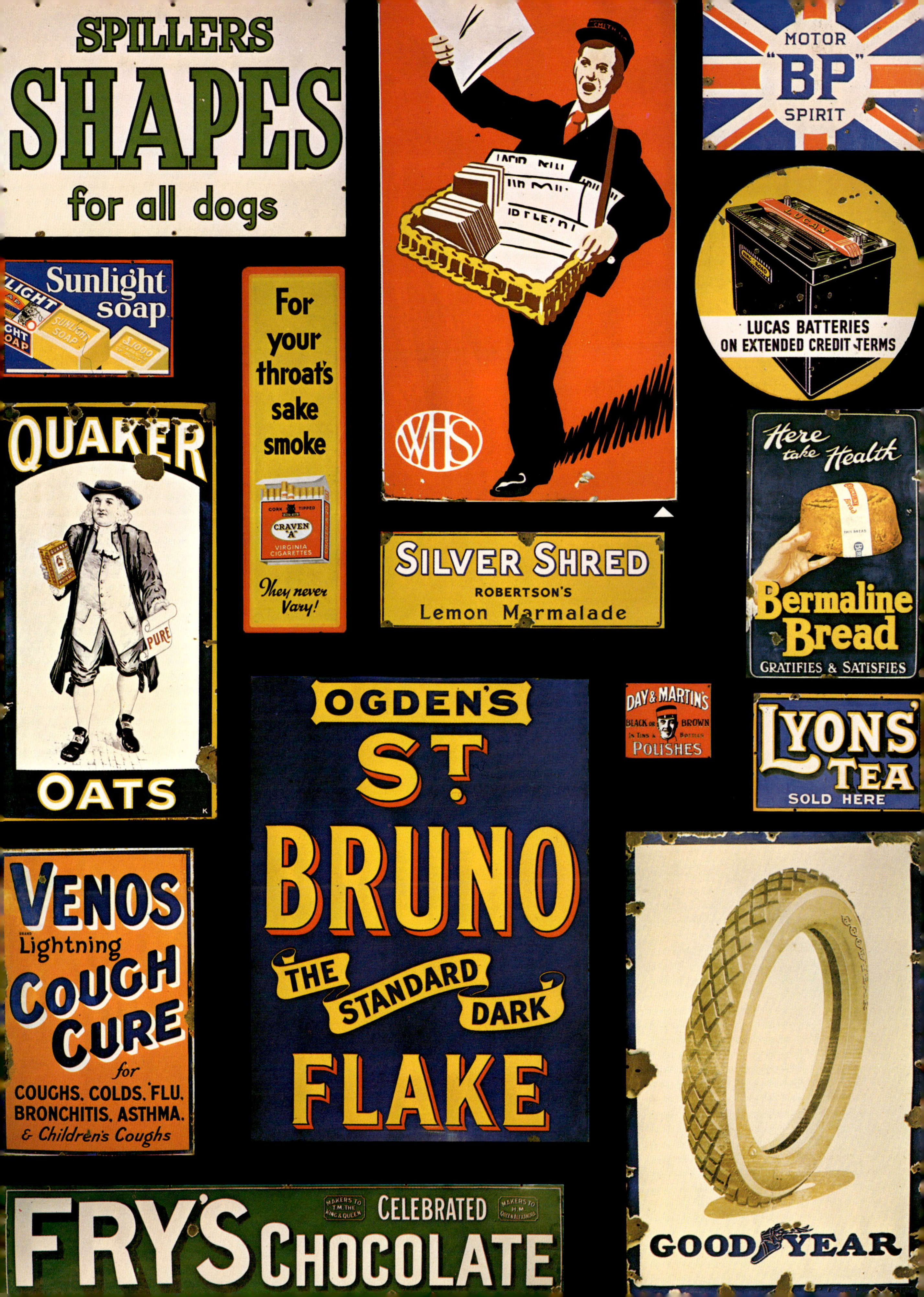

ST. JULIEN TOBACCO

OGDEN'S ST JULIEN TOBACCO
EXQUISITE COOL and FRAGRANT
PURE Blended Virginia

BIBBY'S "CREAM EQUIVALENT" The Best of All CALF FOODS.

1d DAILY MIRROR 1d
BEST ALL ALONG THE LINE

GOLLIBERRY
ROBERTSON'S BRAMBLE SEEDLESS
A Perfect Jelly Preserve

Brooke Bond Tea

USE BURNARD & ALGER'S Celebrated Concentrated "CORN" & GRASS MANURES
PLYMOUTH.

FRY'S CHOCOLATE
300 PRIZE MEDALS
by Royal Appointment

THE POWER BEHIND THE LENS
SELO FILMS
SELO MADE IN ENGLAND
FILMS DEVELOPING & PRINTING

BRITISH MANUFACTURE
ASK FOR TAYLOR'S Celebrated WITNESS CUTLERY
Pocket Knives, Razors, Scissors
Table Knives & Silver Plate
MANUFACTURED AT EYE-WITNESS WORKS, SHEFFIELD ENGLAND

"WILD WOODBINE"
"WILD WOODBINE" CIGARETTES
W.D & H.O WILLS BRISTOL & LONDON
CIGARETTES

Banish CONSTIPATION overnight with BROOKLAX

DISPENSING CHEMIST DRUGS | TOILET ARTICLES FILMS
THE BRITISH
BROOKLAX
CHOCOLATE LAXATIVE

2d each "RAJAH" CIGARS SOLD HERE

PIONEER CEMENT
QUALITY UNRIVALLED
MANUFACTURED BY CASEBOURNE & Co Ltd
Haverton-Hill-on-Tees

Daily Telegraph

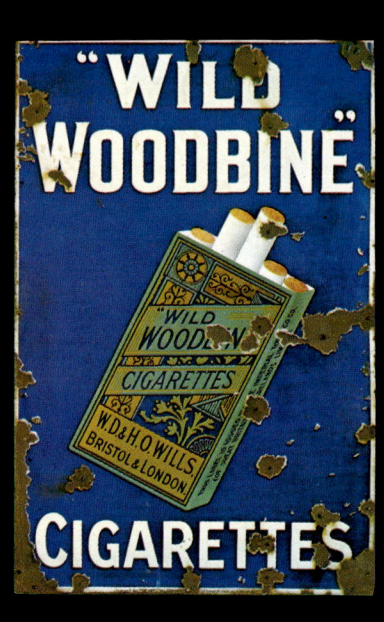

Rowntree's CHOCOLATES

LYONS' INK

VOLVOLUTUM
NO RUBBING! NO SOAKING!!
Mrs 'V' SOAP

OXO
OXO CUBE
IT'S 'MEAT & DRINK' TO YOU
MELOX DOG FOODS
"TURF"
"TURF"
VIRGINIA CIGARETTES
20 CIGARETTES 20
VIRGINIA CIGARETTES
GUARANTEED MADE FROM TOBACCO AGED AT LEAST 3 YEARS IN THE WOOD
Britax
SAFETY BELT FITTING SERVICE
BATTLEAXE BAR
OGDEN'S WALNUT PLUG
SWEET AS A NUT
BOVRIL
"Oh Mamma don't forget to order BOVRIL"
Smoke
PLAYER'S NAVY CUT
Tobacco and Cigarettes
SHELL
CARS FOR HIRE
ELLIMAN'S
EMBROCATION
YOUR TOBACCONIST
ZEBRA
Grate Polish
THE SMOKER'S MATCH
SWAN VESTAS
BRITISH MADE BY BRYANT & MAY
POWERFUL
HUDSON'S SOAP
EASY AND SAFE
VENUS SOAP
GUINNESS IS GOOD FOR YOU
GIVES YOU STRENGTH
You know it by this.
"HIS MASTER'S VOICE."
SPILLERS
Cattle, Pig & Poultry Foods

DUCKHAM'S OILS
PERFECTLY REFINED
POLISH DE LUXE METAL KOMO PASTE
ENTIRELY FREE FROM GREASE
A PERFECT POLISH ALSO FOR MIRRORS
"KOMO FOR METALS"
Milkmaid BRAND Milk
Largest Sale in the World
TRADE MARK
OIL FILTERS AC PLUGS
GUINNESS IS GOOD FOR YOU
GIVES YOU STRENGTH
FRY'S BREAKFAST COCOA
MAKERS TO H.M. THE KING. H.M. THE QUEEN
H.M. QUEEN ALEXANDRA.
FRY'S PURE BREAKFAST COCOA
300 GRANDS PRIX, GOLD MEDALS &c.
J.S. FRY & SONS, LD.
BRISTOL & LONDON
4½D. PER ¼ lb TIN 4½D.
GOODYEAR
PLAYER'S
PLAYER'S 10 Cigarettes "DRUMHEAD" CIGARETTES.
"DRUMHEAD" CIGARETTES
spend wisely
Brooke Bond dividend Tea
save wisely
WALTER WILLSON'S LTD
Smiling Service Shops
TRADE MARK
For Better Food
DIAMOND ALE
NORTHAMPTON
REDBREAST
OGDEN'S REDBREAST Flake
ROUGH CUT
FLAKE
VENO'S Lightning COUGH CURE
Cures COUGHS, COLDS & CHEST DISEASES.
OVUM
THORLEY'S POULTRY SPICE
SHELL

MARTINI
MARTINI
MARTINI
VERMOUTH
NESTLÉ'S
ONE QUALITY ONLY
RIGHEST IN CREAM
MILK
CARROLLS
Number 1
TIPPED
DRINK
Coca-Cola
IN BRANDSTOFFEN
DRINK
Coca-Cola
STRIKE MATCHES HERE
Pepsi-Cola
ESO ES!
EL REMEDIO
COOPER
NO HAY IGUAL
BELGA
VANDER ELST
CLEVELAND
PETROLS
VAN HOUTEN'S COCOA
RENAULT
GOOD YEAR
GOOD YEAR
TYRES
GARGOYLE
REGD TRADE MARK
Mobiloil
VACUUM OIL COMPANY LTD
SPA
VAN
a cocoa you can enjoy
HOUTEN'S
COCOA
BEST & GOES FARTHEST

GUINNESS
is good for you
REDGATE
SOLD HERE
Table Waters
Enjoyed the year round since 1878!
MAKERS TO H.M. THE QUEEN
Rowntree's
PASTILLES
MORRIS
Sales and Service
"MATCHLESS"
METAL POLISH
Makers by Royal Warrant to Her Majesty the Queen
JONES SEWING MACHINES
LISSEN
MICA PLUGS
GUARANTEED FOR ONE YEAR
ADKIN'S
NUT BROWN
TOBACCO.
OGDEN'S
WALNUT
OGDEN'S
WALNUT PLUG
SLICED
PLUG
"ACE OF SPADES"
SPECIAL BRAND
LUCAS'
SOLD HERE
Colman's
Starch
NECTAR
TEA
SMOKE
PLAYER'S
"AIRMAN"
TOBACCO
Drink
TIZER
The Appetizer
Coolness, Fragrance.
A MEDIUM MIXTURE
"The Chairman"
R.J. LEA Ltd
MANCHESTER
Distinctive Flavour.
6d Per Oz.
MAURETANIA
HOLZAPFELS
COMPOSITIONS

HALL'S DISTEMPER
For Healthy Homes
SISSONS BROTHERS & Co Ltd
HULL. & 199a Boro' High St London. S.E.
Liverpool Glasgow Manchester Reading

GUARANTEE OF PURITY
£1000
SUNLIGHT

"Black Cat"
Pure Matured
Virginia Cigarettes
Carreras Ltd. (Estd 1788) London, England

REMOVALS
AND
STORAGE
BY
ROBSONS Ld
42 NORTHUMBERLAND STREET

REMOVING & FURNISHING
FURNITURE
STORED
J. LENO & SONS
UNIVERSAL STORES
UXBRIDGE.

DELICIOUS
MAZAWATTEE
TEA

ASK FOR
DARTNELL'S
PRIZE MEDAL
A. DARTNELL
BREWED GINGER BEER
NORTHAMPTON
GINGER BEER

SINGER
SEWING
MACHINES

COOPER
COOPER'S SHEEP DIPPING POWDER
THE STANDARD SHEEP DIP

FURNITURE
BOUGHT
FAMILIES REMOVING
OR
WAREHOUSING
FURNITURE, LUGGAGE, &c
TERMS POST FREE.
TAYLOR'S DEPOSITORY
RANELAGH ROAD,
PIMLICO, LONDON.
SOLD
AND SHIPPED

1D.
MONSTERS

Veritas
INCANDESCENT
MANTLES
SOLD HERE

Anti-Laria
A Non-Alcoholic
SPARKLING
WINE
A STIMULANT SECOND ONLY TO CHAMPAGNE
AT A FRACTION OF THE COST

OGDEN'S
COOLIE CUT
PLUG

STRANGE'S
A-1
CRYSTAL
OIL
REGISTERED
TRADE MARK
LIGHT
The
Finest & Safest Oil
Sold in the WORLD
"WHITE MAY"
AND
"ROYAL STANDARD"
"BP"
LAMP OILS
TURF
TOP QUALITY CIGARETTES
DON'T BE MISLED !!!
DRINK
"CAMP"
IT'S THE BEST
Sole Proprietors R. PATERSON & SONS LTD. GLASGOW
INGERSOLL
INGERSOLL
WATCHES
5/- & UP
FOR SALE HERE
OGDEN'S
BATTLE AXE
OGDEN'S
BATTLE AXE
SLICED BAR
BATTLE AXE
SLICED
BAR
BURMA
SAUCE
BURMA
SAUCE
THE ONLY "SAUCE"
I DARE GIVE FATHER
GET YOUR
Player's
Please
HERE
COLMAN'S
D.S.F.
MUSTARD
£1000
SUNLIGHT
SOAP
GUARANTEE OF PURITY
THE ORIGINAL
STURMEY-
ARCHER.
3 SPEED &
"TRICOASTER"
FITTED HERE
COOLIE CUT PLUG
FULL FLAVOUR
SWEET & LASTING
WILLS'S
"FLAG"
FLAG
CIGARETTES
W. D. & H. O. WILLS
FLAG
CIGARETTES

WILLS'S
GOLD FLAKE CIGARETTES
10 CIGARETTES
W.D.&H.O.WILLS, BRISTOL & LONDON.
CIGARETTES SOLD HERE.
BROOKE, BONDS' TEA.
SMOKE
CRAVEN "A"
WILL NOT AFFECT YOUR THROAT
will not affect your throat
SMOKE
WALNUT PLUG
"Sweet as a Nut"
WE SELL
ELEY KYNOCH
CARTRIDGES
LYONS TEA
Drink
TIZER
The Appetizer
COLMAN'S MUSTARD
THOUSANDS IN USE
PETTER
OIL ENGINES
& ELECTRIC LIGHTING PLANTS
WORKS:- YEOVIL, ENGLAND.
SPECIAL
NOSEGAY
MAKES A SPECIAL CIGARETTE
HUNTLEY & PALMERS
GINGER NUTS
John Ginger
READING & LONDON.
SWAN
WHITE FLOATING
SOAP
ABSOLUTELY PURE
GET
PRATT'S
HERE
C.W.S
LIQUID METAL POLISH
FOR ALL METALS
PELAW
SOLD BY ALL CO-OPERATIVE SOCIETIES
Brilliant Lasting
Economical Easy
PELAW
LIQUID
METAL POLISH

HUDSON'S SOAP
FRY'S
CHOCOLATE
WILLS'S GOLD FLAKE
CIGARETTES-10-3
DRINK
"CAMP"
IT'S THE BEST
CHURCHMAN'S
"TORTOISESHELL"
SMOKING MIXTURE
CADBURY'S
CHOCOLATE
Delicious Wholesome
Always ask for
CADBURY'S
MAZAWATTEE TEA
SWAN VESTAS

Lyons
Cooling News
WE SELL LYONS' ICE CREAM
Lyons
$2\frac{1}{2}$D FRUIT PIES
MANY VARIETIES
WATSON'S MATCHLESS CLEANSER
SOAP

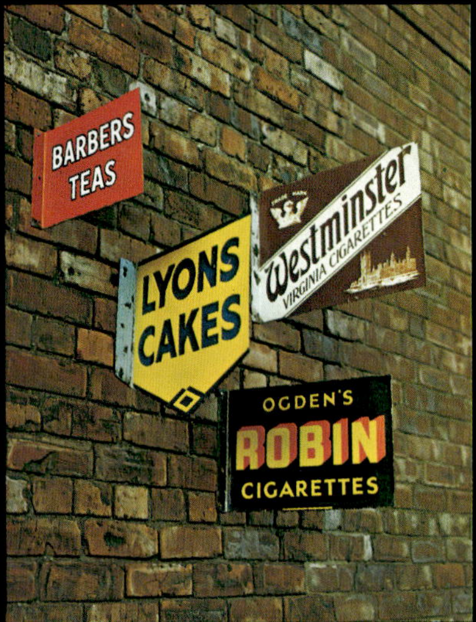
BARBERS TEAS
LYONS CAKES
Westminster
VIRGINIA CIGARETTES
OGDEN'S
ROBIN
CIGARETTES

GUARANTEED
SHELL FROM THE PUMP
SHELL lubricating OILS
FROM THE PUMP SHELL
SHELL
MOTOR OIL
MOTOR
"BP"
SPIRIT

Design, Texture, Typestyles and Stencil Patterns

DESIGN

An examination of the major design elements such as size, shape, colour and content, provides a clear insight into the developing taste of succeeding generations of advertisers.

Sizes ranged from as little as a few inches square, to composite giants covering areas in excess of forty square feet. An example of the former is Veritas Gas Mantles and the latter Van Houten's Cocoa. However, the average size of signs was from three to four feet in height and width, with occasional long and narrow expanses of six or seven feet by a few inches. The factor determining this as an average size was generally the restriction of locations, as already indicated, and the majority of signs were to be found underneath shop windows. However, other locations on shops themselves, dictated different sizes and proportions such as shapes that would fit vertically on door frames and horizontally on counter fronts.

There were two main shape categories, those that were not based totally on the particular product being advertised and those cut out in the form of the product that the sign was advertising. Of the latter type, Nectar Tea was in the shape of a cup and saucer, Mobil and Shell Oils both took the form of cans and Mazawattee tea was delicately cut out around the saw-tooth shape of a tea leaf. Some shapes, though not made in the form of the product, were cut around intricate outlines. For example, figures such as the Sunlight Soap boy offering a bag of £1,000 reward money perched on a bar of soap, or the man riding a Raleigh bicycle. Of the first category, apart from the normal rectangular shapes, other general shapes could be seen in the form of ovals, circles, triangles, lozenges, arrows, shields or a combination of these. Examples of these were: oval – Nestlé's Milk, circle – Lyons' Cakes, Lucas Batteries, Cooper's Sheep Dip, triangle – BP and Shell Oils, lozenge – Renault Cars, Vantas Lemonade, John Bull Tyres, arrow – Shell Petrol, shield – BSA Bicycles, Walter Willson's Stores, combined shapes – Craven 'A' Cigarettes, Karpol Car Policy, Morris Cars. It must be added that many signs were double-sided, projecting from the wall and attached by brackets, chains and hooks. This method not only doubled the visual effectiveness of the sign, but also added to its strength and stability, since there was thick enamel on both sides. Normally only a thin layer of counter-enamel was put on the reverse side during processing. Examples of these double-sided signs are: Royal Daylight Lamp Oil, Will's Star Cigarettes, Lyons' Cakes, Shell Lubricating Oil, Hay's Cleaners, Churchman's Number 1 Cigarettes. Some, like Lyons' Cakes signs had elaborate wrought iron frames. Domed or bulbous signs used for early Lyon's Tea and late John Bull Tyres were never popular in this country, but they were used extensively on the continent, in fact, the original Lyon's tea signs were ordered from Germany. Another example of this type is the Fry's Cocoa Shield.

For the convenience of the customer, stores would often provide chairs to sit on while they gave their orders and often these chairs were issued to the shopkeeper by the advertisers who had had inserted into the back of the chair an enamel plaque such as Watson's Matchless Cleanser or Venus Soap. Other three-dimensional signs took the form of 'A' shaped sandwich boards which stood on the pavement outside shops and were portable and free-standing. Companies using this method were Palethorpes, who had a set with sausages on one side and a pork pie on the other and Lyon's who had a fruit pie on one side and an ice cream advertisement on the other.

There were also enamel plated vending machines attached to walls, advertising the contents such as chocolate, mints, razor blades and cigarettes.

Unframed signs were usually large and found at a high level above a fascia or on gable ends. Framed signs normally consisted of a plate, slotted or screwed into a wooden moulding, sometimes strengthened with angle plates on the back at each corner. On long signs it was often necessary to brace these with cross stretchers. The whole sign and frame was fixed to the building fabric by means of drilled plates or alternatively screwed directly through the face of the frame. The techniques of fixing varied from one sign manufacturer to another, but in most cases the fixings were supplied painted gloss black. However, most framing, painted or otherwise, naturally rots with the passing of the years.

Unframed signs were usually attached by means of either screws and lead washers, through ready-made holes in the sign, or masonry nails with flat heads, retaining the sign where the holes were either inappropriate or not included. Most signs, from a few inches square up to six feet square, had between two and eight screws or nails per side. Pointing was occasionally used as an additional means of support on uneven wall surfaces. These unframed signs tended to corrode faster at the fixing points due to the total exposure of the edges and back. Points of corrosion also arose from the stone and air gun attacks of generations of youths.

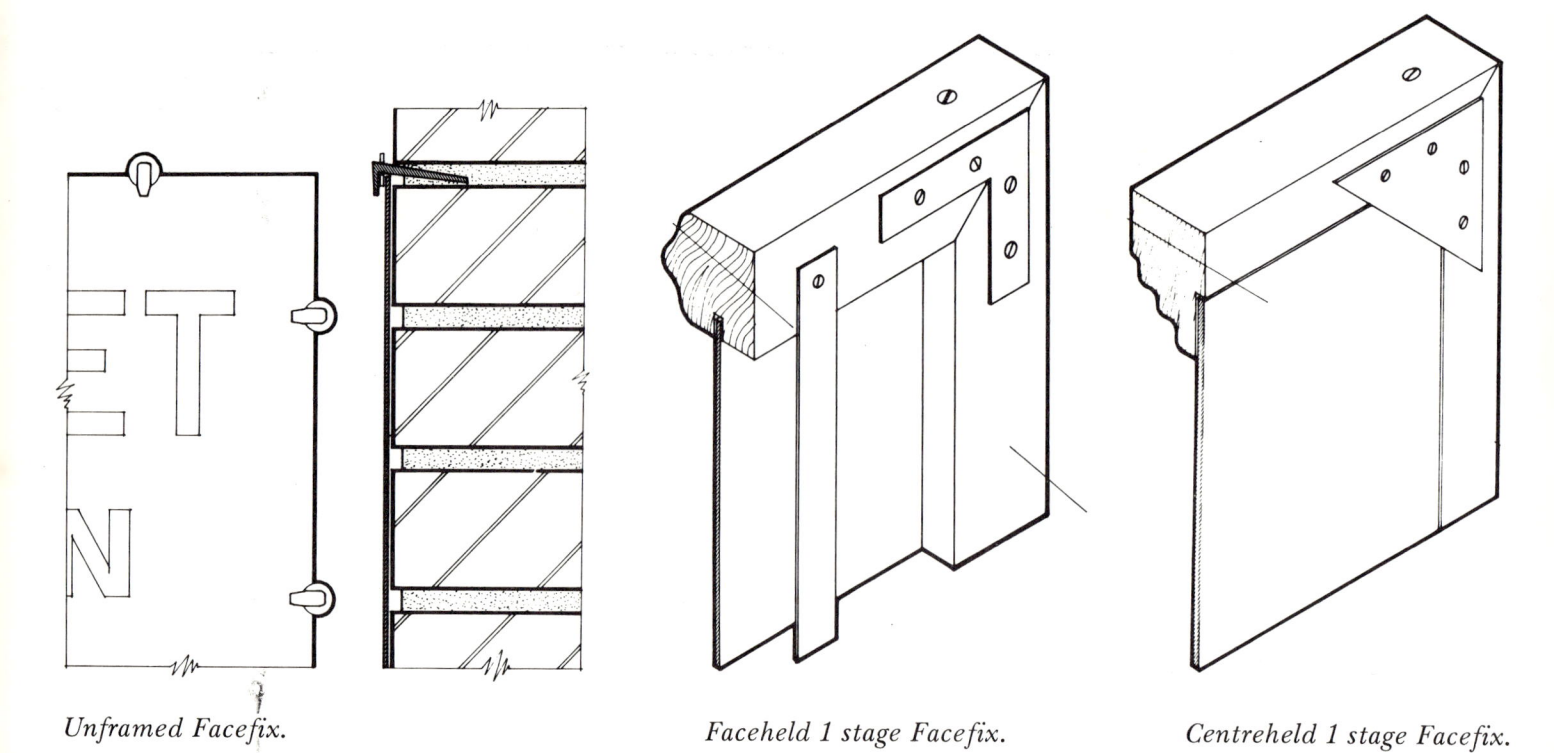

Unframed Facefix.

Faceheld 1 stage Facefix.

Centreheld 1 stage Facefix.

52

TEXTURE

The area of heat-fused glass from which the surface of an enamelled iron sign is created has certain textural and visual qualities peculiar to the enamel medium, which are particularly pleasing and in sharp contrast to the surface of other pictorial media such as paint, wood, paper or plaster. Each colour is individually applied — on top of or slightly overlapping the edge of the one below — and each has its own distinct edge separating it from its neighbours. A meniscus is formed during the period when the glass is molten and remains on the edge of each layer and this can best be appreciated by running the fingers gently over the undulating surface. Some manufacturing methods created a more exaggerated manifestation of this effect, so that the surface of the Ogden's series, with blue background and red and yellow lettering, is particularly bumpy. The play of light over the surface is broken up into variegated planes and facets, which are enhanced if the sign is in good condition and carefully polished. Enamel is semi-transparent and highly reflective and can produce colours and tones of great richness and depth. The frequent use of bright primary colours and the sparkling bright surface effect that the best designs afford, linked with the similarity in materials used in fine and applied art enamelling, suggest the simile 'street jewellery'.

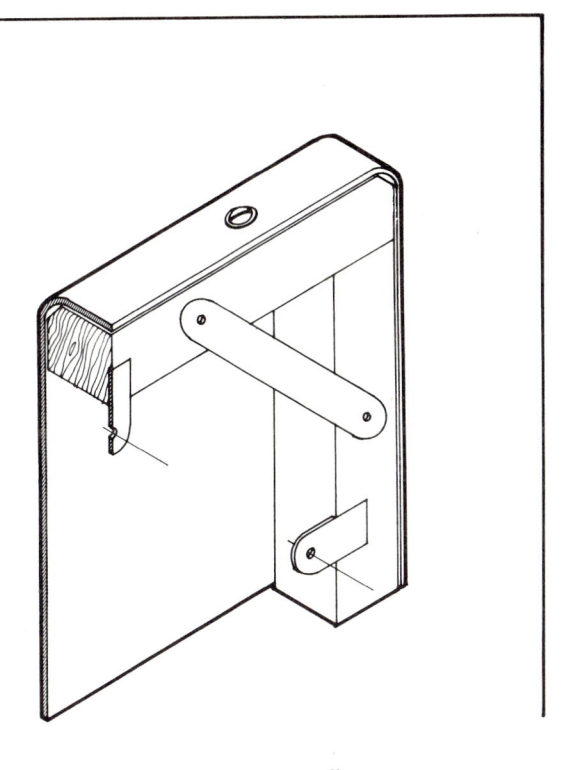

Rearheld 2 stage Secretfix.

TYPESTYLES

One of the most appealing characteristics of enamelled advertising signs, from a pure design and historical aspect is in their embodiment of a variety of differing typestyles. Each one was chosen to perform its own eye-catching function.

It is interesting to note the use of free-flowing lettering for Rowntree's pastilles, the italic script of Veritas Mantles and the soft, three-dimensional forms used on Player's Airman and Camp Coffee. Our eyes pass on to the harder forms of block shadows on the serif letters of Shell and to the sans-serif of Tizer and outline on Raleigh.

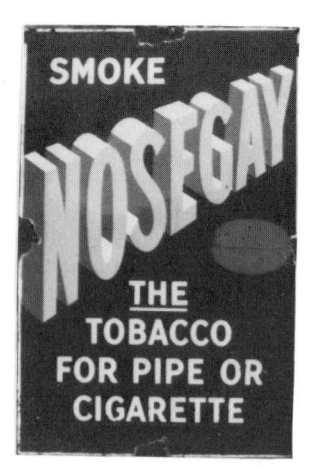

54

All these typestyles and their variants help us to date signs as well as giving each its peculiar individuality. Several signs illustrated show the use of particularly fashionable typefaces of certain eras. For example the Art Nouveau period is represented by the ornate lettering of Swan Ink, Lucas Spades, Nectar Tea and Bovril.

Another interesting aspect of letter forms on signs is the use of a number of different styles on the same plate. Palethorpes use three completely different faces in as many words. Similar richness of styles can be noted on P & R Hay Dyers & Cleaners, and on the brown background of Fry's Chocolate.

LYONS'
TEA
SOLD HERE

THOUSANDS OF PEOPLE EAT
Bermaline
Bread
just because it is nice - But what a
lot of good they are doing themselves
in a quiet way!

REDGATE
Table
Waters
ON SALE HERE
BEST SINCE 1878
REDGATE LTD. TRAFFIC STREET, NOTTINGHAM.

TADDY'S
IMPERIAL
TOBACCO

WAVERLEY
CIGARETTES

AVON
TYRES

STENCIL PATTERNS

The earliest method of applying a pattern to the iron plate was by the elaborate and skilful use of stencilling. Where more than two colours were used, a considerable feeling for the essence of the medium was required of the designer. To build up stylised images by means of multiple stencil overlays, without the use of half-tones, was a highly skilled job and required a bold sense of pattern and tonal appreciation, for a successful result to be achieved. With the introduction of lithographic techniques as an imitation of and in parallel use by other poster-type advertising media, a basic honesty and integrity in the medium was lost. Even when the image was produced by transfer or litho printing, the major colour areas and the larger lettering were usually stencilled, e.g. the rare early photograph and stencilled Royal Daylight American Lamp Oil. But a comparison between an image produced exclusively by stencilling – Champion Norfolk Boots – and one made by borrowing a printing medium – Holzapfel's Compositions – which is a monochrome lithograph tinted with stencilled areas of colour, indicates the purity of the one and the decadence of the other. The intrinsic quality of the enamel texture is still retained in both cases, making both varieties visually and technically acceptable to the collector.

Screen printing was first used by the National Sign Company, circa 1923, and here the effect of stencilling, litho and photography could be combined in one process, e.g. Spillers Balanced Rations, made in Newcastle. Nearly all the modern enamel signs are made using the screen printing process and applied on steel. A direct comparison between an original sign of the early period and a modern reproduction gives a clear demonstration of the difference in quality and finish between the old process and the new. A certain clarity and luminosity typifies the originals, whereas the modern versions tend towards the clinical and meagre.

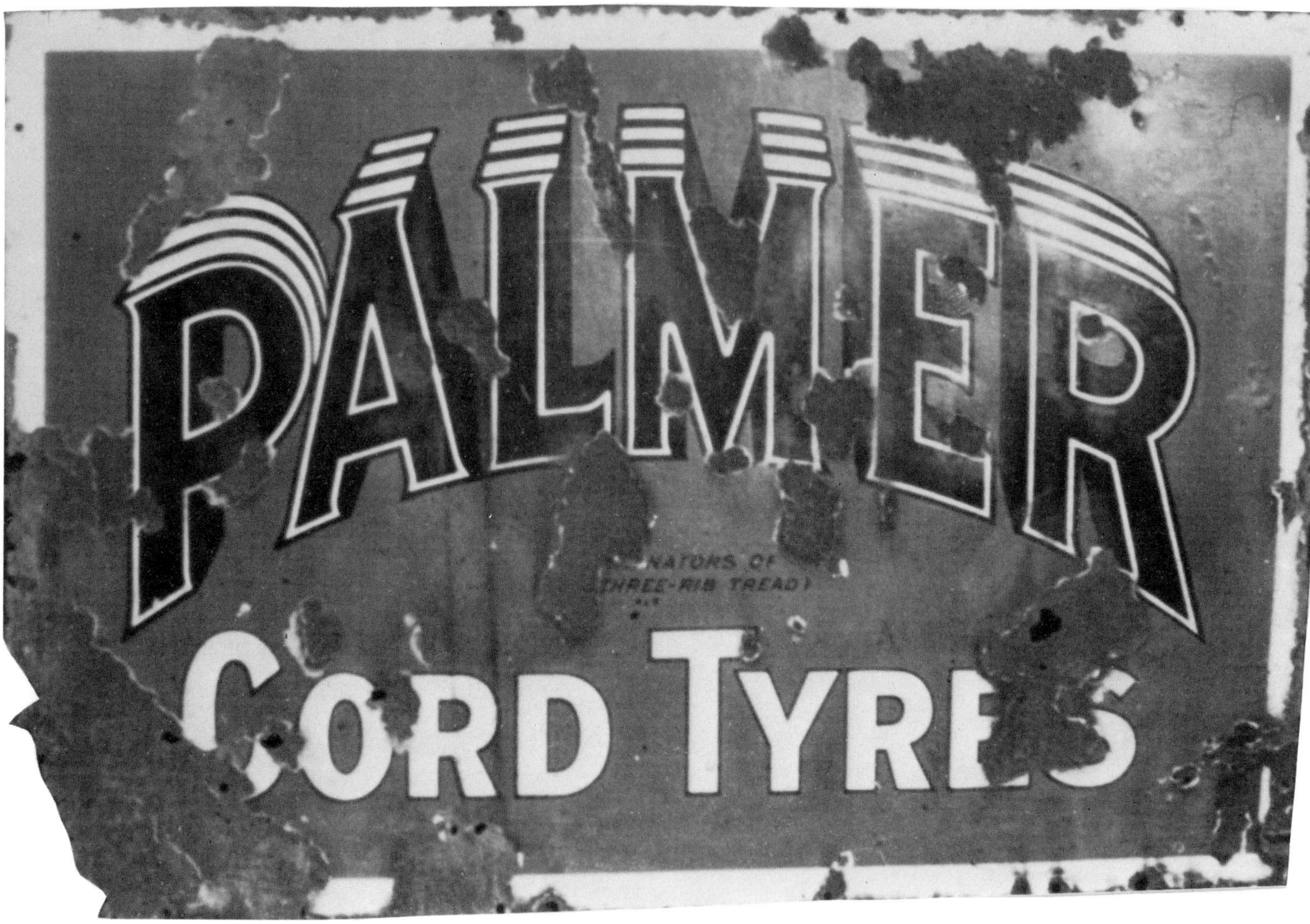
PALMER
CORD TYRES

Collecting, Conservation, Restoration

COLLECTING

Not all collectors of enamels concentrate exclusively on signs. Frequently a sprinkling of enamel advertisements will usefully augment another interest. Thus, for example, for some people the prime consideration is the product shown on the sign. Motoring enthusiasts have dozens of of examples to choose from, though not many show actual vehicles. Sun Motor Insurance is one and another is a fine image of Malcolm Campbell and the Bluebird on a Castrol sign. In other spheres of interest, specialists will consider signs illustrating bottles, packages and tins to have premium value. Also included in the latter category are those signs displaying out-dated prices for products, e.g. Whitbread Pale Ale 2/6 a dozen, and out-moded trading slogans.

Specialisation by the collector in certain product types or brands will affect the value of signs. The parallel trends in ephemera collecting, such as labels, bottles and packaging often influence this. The signs featuring bottles – Dartnell's Ginger Beer, Green & Leddicott's with Zeppelin bottles and Brighton Pier, soap and cigarette packs – Hudson's, Sunlight, Woodbine – or labels and trademarks – Guinness and Gossages – will have special preference and enhanced value for collectors specialising in examples of these items themselves. Similarly, certain non-advertising enamel signs hold special interest for collectors of railway or military ephemera, such as station nameplates or recruiting notices.

Since no sign is more than a century old, and most signs of the last thirty years are not much admired by collectors, the parameters of vintage for individual signs will fall into the last two decades of the nineteenth century and the first of the twentieth. All signs thereafter may be classified as modern. Victorian objects of all sorts have a special prestige for collectors of advertising ephemera and in this respect, enamels are no exception. However, apart from this consideration, the relative age of a sign is not a strong determining factor. Some designs remained in production for decades and as such are difficult to date precisely. But, as the sylisation of typefaces and imagery, typical of Victorian and Edwardian manufacturer often finds a more appreciative audience than those of the inter-war years, so these older examples, (provided they combine some of the appropriate qualities of bold and variegated typestyles, quaint imagery or illustration, an out-moded message or a product with bright colours and good condition) tend to be the most desirable and hardest to obtain. It is now extremely rare to find this type of sign *in situ*. Those that do survive are usually in collections already or turn up in the storage areas of old business premises. Signs with similar visual qualities, but of more recent date, will fetch comparable prices. Again the desirability of signs with less visual richness i.e. lettering only, a few colours and no image, depends on content rather than on age. Only if a date is prominent on the design e.g. Sunlight Soap Highest Award Chicago 1893, or a reference to a monarch (ibid) will the value of the piece be influenced. Sometimes a date code can be

60

WAY OUT

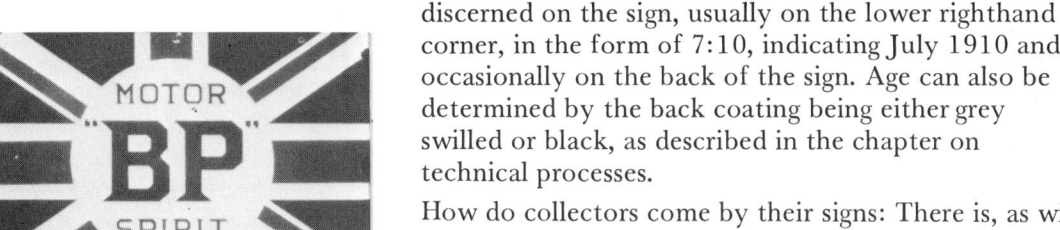

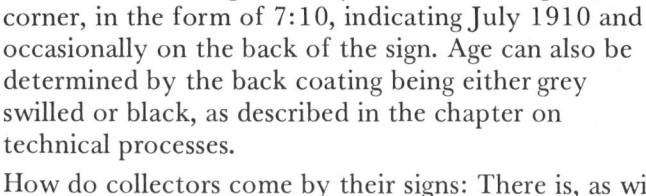

discerned on the sign, usually on the lower righthand corner, in the form of 7:10, indicating July 1910 and occasionally on the back of the sign. Age can also be determined by the back coating being either grey swilled or black, as described in the chapter on technical processes.

How do collectors come by their signs: There is, as with any collectable, a floating market exchange between dealers and collectors, in antique shops, flea markets, auto jumbles and especially from private specialist dealers. However, most big collections are created by the patient scouring and searching of likely and unlikely locations.

In the normal course of events the locations and usefulness of these old advertisements vanish with time and custom. Thousands of miles of terraced streets, which once served as display areas for a thousand or more signs, have disappeared since the War years. Sometimes these signs were systematically removed by their original distributors, but more often they were simply abandoned. The big tobacco firms and petrol companies were particularly active in pursuing a policy of constant renewal of their advertising images, and consequently they replaced and updated their signs. The remainder were left, unmaintained, on the walls. From this abandoned wealth of material only a small proportion has survived and been preserved in good condition, either accidentally or by intent, thus some examples are quite rare now and are sought after by collectors and museums.

The factors determining the collectability of signs and their consequent monetary value, vary according to circumstances. With a mass-produced object such as a sign, rarity is hard to establish, as occasionally large *câches* of particular signs are discovered in warehouses where they have lain, well preserved, for years. Even under the quite arduous conditions of chance survival, such as re-use for building material in a garden shed or compost heap, or buried in a rubbish dump, the enamelled iron sign can survive remarkably well.

61

One factor which for most serious collectors underlies all others, is the condition of the sign. A mint sign, or one with minimal damage such as flaking of the enamel around the screw holes, or slight chipping or corrosion around the edges, is highly prized, as with all collectables. But if it is visually unexciting, even a mint condition sign is not desirable. The most perfectly preserved old specimen of a two-colour, simply-lettered and phrased sign, with no illustration and advertising a dull and mundane product, will not enhance a collection. It is easy to understand the value put on illustrations, and of conveniently small size, say 3' square or smaller, lots of bright colour, or lettering that is complex, ornate or stylised.

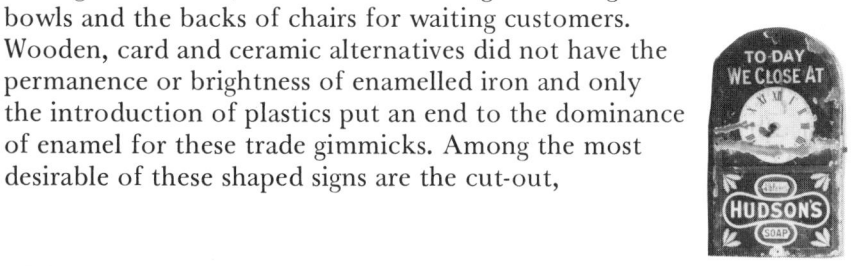

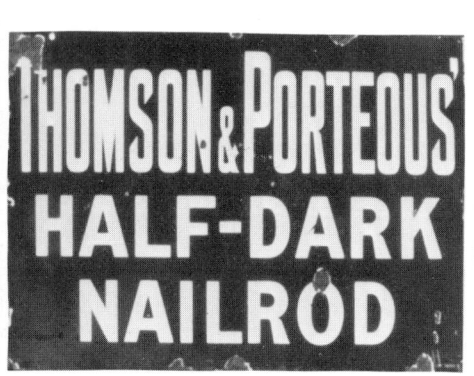

Other desirable qualities in enamelled advertisements are more peculiar to this medium and are not evident in other poster-type advertising. The shaped sign, either geometrically varying from the rectangle or made in a figurative shape for outside use, in practical terms, possible only in enamelled metal, as are opening and closing time clocks, front-of-counter dog's drinking bowls and the backs of chairs for waiting customers. Wooden, card and ceramic alternatives did not have the permanence or brightness of enamelled iron and only the introduction of plastics put an end to the dominance of enamel for these trade gimmicks. Among the most desirable of these shaped signs are the cut-out, projecting, double-sided signs, two examples of which are a pair of hands holding a bottle of 1d Monsters Pop and a fleecy ram extolling the virtues of Cooper's Sheep Dip. The collector can enrich the range of his collection by finding enamelled vending machines, door finger plates, thermometers and barometers. Classics of the silhouette cut-out are illustrated in the colour plates. The Sunlight Soap boy, the Mazawattee Tea leaf are there, but not, unfortunately, the delightful Fremlin's lager elephant.

The main location of *in-situ* signs is still the shop front in the older quarters of industrial cities, either on existing traders' premises or on derelict buildings and gable ends. Generally permission is easily gained from shopkeepers, landlords and estate agents, or original manufacturers for the removal of the signs. It is advisable to offer to undertake remedial work for unsightly or inconvenient gaps left by their removal. This may involve painting, pointing and re-timbering, but for a prize specimen the effort involved is well rewarded.

Removal of the sign itself is a skill only acquired through experience. As the iron/steel and glass of the artefact are brittle and at the same time strong and heavy, controlled strength is required and to climb 40 feet up a ladder means a strong head for heights too!

Equipment required for on site collecting:

Van with roof rack and extendable ladders
Saws for metal and wood
Brace and bit with metal drills to drill out screws
Heavy duty screwdriver
Crowbar and/or levers
Hammer, pincers, pliers
Strong gloves
Penetrating oil and sharp eyes

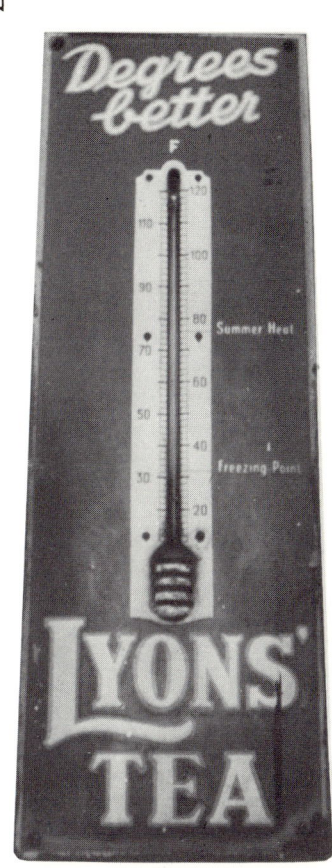

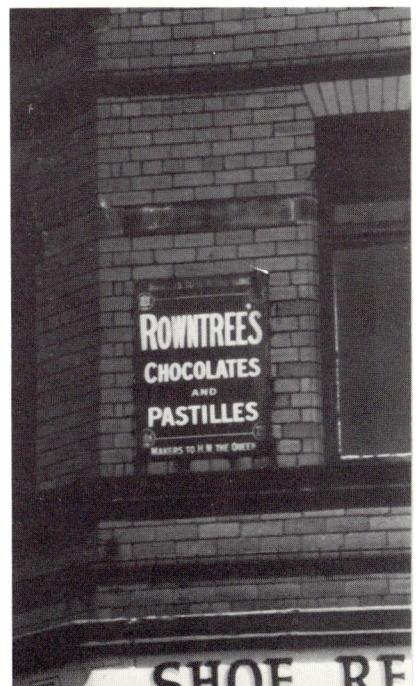

63

CONSERVATION

York Castle Museum.

In the conservation of inner city areas, architects are constantly faced with old forms of advertising on shop fronts, interiors, hoardings and gable ends, which are not always considered worthy of restoration along with the rest of the building. The most common external advertising of any great age surviving today is the enamelled iron sign. These brilliantly coloured signs have endured more than other types of ephemeral advertising because of their strong construction materials and permanent means of attachment. Unfortunately, in new building work on existing property, such signs are far too often relegated to the contractor's waste skip and thus lost to future generations. Enamel signs were used by advertisers because of their special permanence, and thus displaying a strong historical continuity. Their shape was frequently dictated by the facade details of the shop and thus became a decorative architectural feature in much the same way as glazed ceramic tiles and engraved glass windows. They survive as historical documents. If the sign is to be conserved *in situ* then it should be removed for inspection and cleaning. The back will usually have corrosion streaks which can be scoured off and sealed with car underseal or polyurethane varnish. If there is a frame, it will probably be rotted and if so it should be replaced. Even if sound or replaced, it should be treated to inhibit future decay. The front surface of the plate can be cleaned as described in the following section and returned to its fixing.

Although most shop fascias on the street are now renovated and modernised, leaving no room for or trace of enamels, a few museums of bygones (see appendix) have on display reconstructions of street scenes of earlier times which include many examples of such signs. There are also abundant photographic records of signs *in situ* and elsewhere to which the enthusiast can refer.

Mixed media, enamel signs and paper posters at Westminster Bridge.

RESTORATION

When a sign is collected it should be checked over for rust damage on the front and back. If it has survived with the enamel intact, a very rare occurence, then no action need be taken apart from cleaning with the appropriate cleanser, i.e. scouring paste and warm water applied with a cloth for mild surface dirt, or wire wool and scouring powder, with frequent sluicings to shift really ingrained dirt. Care must be taken with delicate transfer and lithographed areas, and with certain black lettering and outline overlays. Rusted areas should be cleaned with a rust removing chemical agent, steel wool or emery and immediately sealed or treated with oil, anti-rust paint or underseal. Broken and chipped enamel cannot be satisfactorily replaced, except in small areas on a good metal surface, but even so the process requires complex equipment and technical skill. A method of replacing missing metal and enamel has been devised, using fibreglass and resin or the plastic padding normally used for car body repairs, and painted with enamel colours. Only expert manipulation of these materials yields agreeable results, so that in most cases of surface damage, the scars are best left undisguised and regarded with pleasure as the patina peculiarly associated with the medium. The construction of a wooden frame on which to screw the signs will usually be beneficial, in that a means of attachment to the wall is provided that does not involve danger to the enamel surface caused by repeated screw pressure or hammer blows. The plate is kept rigid and thus free from the splintering caused by bending and torque.

Restoration technique:

1. After thorough cleaning and de-rusting make sure the sign is rigid and flat by fixing with round-headed screws to a wooden frame. Lead or nylon washers may be introduced between the screw head and the enamel to avoid crushing the surface.

2. Back holes with aluminium mesh and resin. A greased card held to the enamel surface will keep bleed-through-resin to the correct level.

3. Fill the hole to the level of the enamel surface, repeatedly scraping it down rather than sanding, to keep the surface smooth and to avoid scratching the enamel.

4. Lightly paint with Keep's enamel paint, avoiding brush marks. Air brushing achieves best results.

NB. Painting in a dust-free atmosphere prevents a grainy texture. Blues are the hardest colours to match. Sometimes a white undercoat will help.

All signs benefit from a vigorous wax polishing.

Burning paint off a sign.

For the social historian the advent of mass advertising serves as an invaluable window through which the life of the late nineteenth century and early twentieth century can be viewed. For earlier periods we are dependent for such insights on those few buildings and cultural and domestic relics that have survived, and which generally reflect the life of only a small section of society.

Early mass advertising, of which the enamel sign is a significant and well preserved part, reflects the needs, desires, and requirements of a large section of industrial society, in particular the working and lower middle classes. It is a primary source of evidence of the fact that by the 1890s these sections of the populace were forming a significantly consumer based society, with an appreciable amount of disposable income. Various products which before the era of mass industrial production would have been luxury products, were now within the grasp of a large section of society. Raleigh, the all steel bicycle, Singer sewing machines, Petter's Oil engines for electric lighting plants are just a few examples of this. Increasing literacy is reflected, not only by the increasing use of the written word in advertising, but also by signs advertising writing materials such as Swan and Stephens inks and MacNiven and Cameron's pens. Increased awareness of hygiene and cleanliness is seen in the emphasis on soap advertisements, and improved conditions of comfort are indicated by the implied ability to purchase carpets – Chiver's carpet soap. Captured in its infancy is that product which so perfectly symbolises the age of mass production and popular affluence – the motor car – Morris, Sun Insurance and petrol.

The current face of advertising is so quickly outdated and superseded that we are accustomed to constant novelty in the medium. Advertising forms, too, are ephemeral. Thus, when a person who has been aware of consumer marketing campaigns for several decades is confronted with an example of advertising in a form that has long since vanished it is often with a feeling of friendly familiarity, as that of meeting a long-lost school chum for the first time in many years, that he will recognise and appreciate the old advertisement.

This syndrome is particularly true of the enamelled iron sign, since when they were first introduced, they remained in being for long periods and became landmarks in a local area. 'Turn right at the church and left at the Woodbine sign' must have been an almost natural instruction at one time. The familiarity and the association of the sign with the products they advertised and the evocation of using these products, raises a wave of nostalgia and sentimental association in the minds of those able to remember enamel signs as an everyday feature of their environment. A packet of Hudson's soap powder may well spark off a recall of washday, with all the associations of the smell of clothes in hot soapy water, the regular sloshing of a dolly stick in the washing and the noisy gossip and clamour of housewives at the wash house and mangling shop. Who, even now, can remain unmoved by the allure of Tizer the Appetizer (and those summers that always seemed far longer and hotter) and Palethorpe's sausages, when due to the prevalence of their respective advertising campaigns these products' trade-names have become synonymous with the products themselves? Fry's Five Boys chocolate is still remembered vividly by the immediate post war generation and the archetypal symbols of the boy's face in different moods still raises delighted grins, even though the Pavlovian association with mouth-watering chocolate can only lead to desperation since the product is no longer available and whether true or false didn't everything taste that much better then! So it is with much of the produce advertised on the signs. Many companies still thrive who used the medium to promote their goods and are prime users of magazine advertising space and television time. Many are now as defunct as the medium itself. It may lie with the present generation, with its championship of conservation and non-expendability to resurrect the enamel sign as a tree-saving, environment enhancing form of advertisement, with built in non-obsolescence, but that is a perhaps wistful thought, too optimistic to entertain seriously.

FRESH
PALETHORPES
TO-DAY
ANDREWS
LIVER SALT
WILLS's
"WILD WOODBINE"
CIGARETTES
Remington
means
TYPEWRITER
Stephens Inks
140
130
120
110
100
90
80
70
60
50
40
30
20
10
0
BLOOD HEAT
SUMMER HEAT
TEMPE RATE
FREEZ ING
ZERO
FOR ALL TEMPERATURES
Stephens' Inks
BRITISH MATCHES FOR BRITISH HOMES
PUCK
MATCHES
BRYANT & MAY LTD
THE FIRST & ORIGINAL
STURMEY-ARCHER
3 Speed Gear And Tricoaster
MAKES CYCLING EASY
BEWARE OF IMITATIONS
FITTED BY ALL CYCLE AGENTS AND MAKERS
FIRST
FOR QUALITY
LAST FOR PRICE.
NEW
HUDSON
CYCLES
THE WORLDS BEST VALUE
EASY
TERMS
6.12 OR 18 MONTHS.
RIDE A
RALEIGH
THE ALL-STEEL BICYCLE
MORRIS
SERVICE
SINGER
SEWING
MACHINES
SOAK THE CLOTHES-THAT'S ALL!
Rinso
SAVES COAL EVERY WASH-DAY
MACNIVEN & CAMERON'S
PENS
ARE RECOMMENDED BY
3050
NEWSPAPERS
JONES'
"Busy"
SEWING MACHINES
BRITISH MADE.
COGSHULTZ
AMMUNITION
SOLD HERE
MATCHLESS FOR THE COMPLEXION
PEARS
THOUSANDS IN USE
PETTER
OIL ENGINES
& ELECTRIC LIGHTING PLANTS
WORKS:- YEOVIL, ENGLAND.
SMOKE
RAJAH
CIGARS
will not affect your throat
NEWCASTLE-UPON-TYNE
CO-OPERATIVE SOCIETY (LD)
REGISTERED OFFICE
103, NEWGATE STREET.
It pays to pay less and fit LISSEN
LISSEN
MICA PLUGS
GUARANTEED FOR ONE YEAR
Wright's Coal Tar Soap

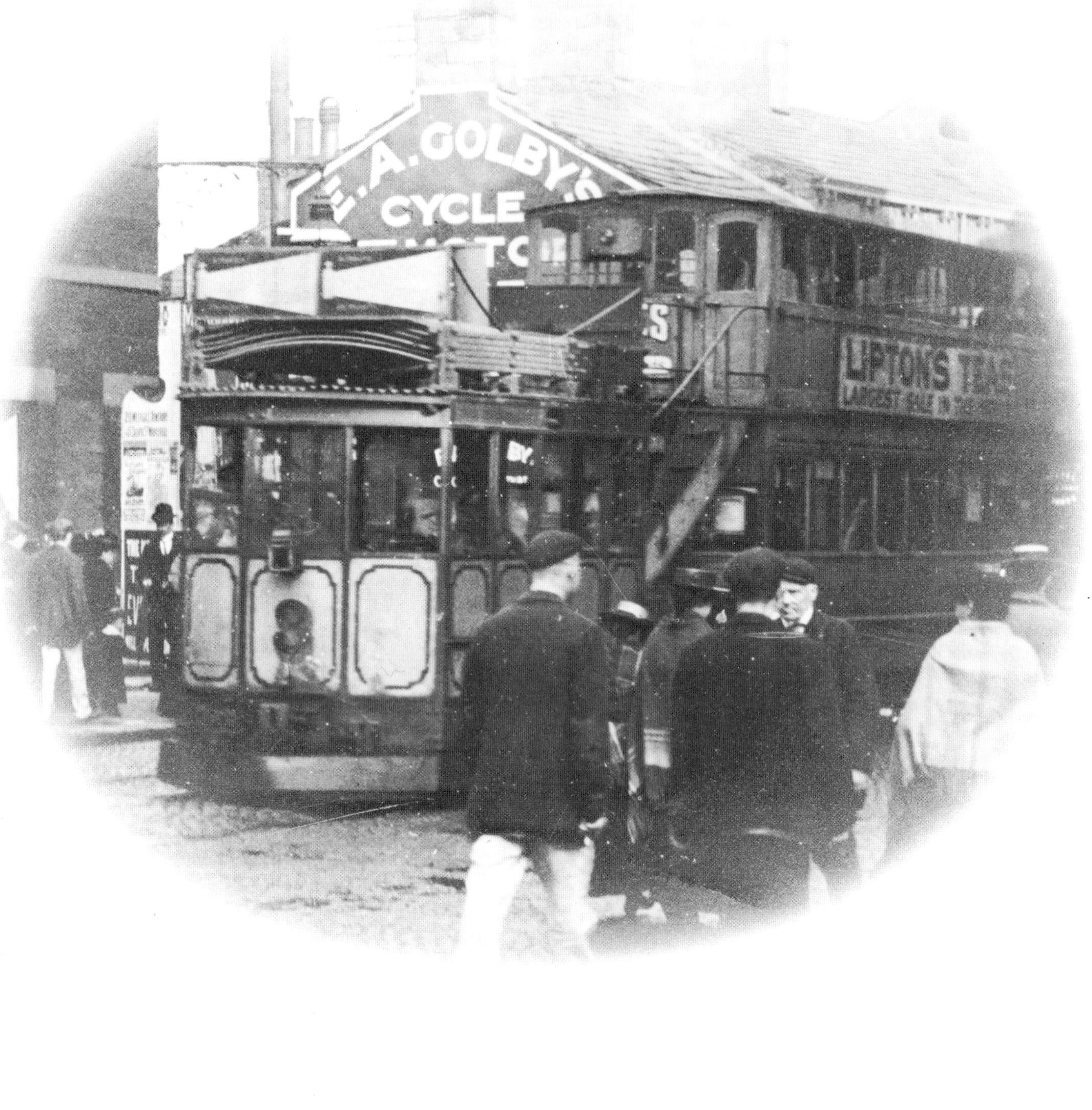
CYCLE
LIPTON'S TEA

Manufacturers past and present

MANUFACTURERS PAST AND PRESENT

Much of the historical and technical data for this book is based on information supplied by Ivor Beard, a director of the late Patent Enamel Co., he being the last of three generations of his family to work there, his grandfather having been a co-founder with Benjamin Baugh. He recalls the hey-day of enamel sign manufacture during his boyhood and how he watched, with regret, the decline in demand which he claims coincided particularly with the end of railway expansion. He recalls several attempts to revitalise the industry by mergers between companies, notably the big three — Patent, Chromographic of Wolverhampton and Imperial of Birmingham. But these measures came to nothing with the onset of the Depression years. Further problems and crises during and after the War included an acute steel shortage, when rationing precluded the use of this vital commodity for advertising purposes. Aluminium which was cheaper and more readily accessable, served as an alternative for a while, but true enamelled iron sign making in the grand scale was gone forever, by the 1950s. Also at this time many of the old enamel sign plants were taken over by manufacturers of baths and cookers for in-factory enamelling facilities. Amalgamations and single company franchise of petrol stations and breweries meant that fewer types of signs were needed and thus orders were denied to the industry. Simultaneously trade was being constantly lost as small businesses, such as ice-cream makers, bakers and soft drinks manufacturers declined in the face of competition from the large combines. The demise of the

Enamelled copper and glass letters made by Burnham's.

small firm and the development of chain stores — although some of these like Walter Willson used enamels to advertise — meant that the small local enamelling firms also went out of business for lack of custom.

One firm which has survived all these crises from its foundation by exiled Frenchman, Charles Garnier in 1890, until the present day, is Garnier Signs of Willesden Green, London and an example of their work is used on the cover of this book. Their factory has always been at Willesden, but up to 1941 their registered office and showrooms were at Farringdon Street, until these were destroyed by enemy action. The Farringdon Street address can be seen on this Lloyd's News sign of circa 1918-23. Other manufacturers who used Garnier signs over the years included Nugget Polishes, Van Houten's Cocoa and Nosegay Tobacco (see appendix) Garnier's made the still familiar, but fast-disappearing, stamped-out enamelled copper letters, which are glued with mastic varnish to shop windows and which advertise Typhoo Tea and Cadbury's. These letters were devised originally by yet another extant firm — Burnham's (Onyx) Ltd. who had a large contract to supply advertising signs for Fry's.

A very extensive use of signs was to plate complete shop fronts in a uniform 'house style'. Garnier's supplied signs for this purpose to Sunlight and Western Laundries and for Boar's Head Tobacco and also to various small merchants and tradesmen who used enamelled plates to cover their trading carts, on fronts, sides and tailboards. Patent Enamel followed a similar course by supplying such overall sign coverage for shop facias to advertise 'The News of the World' — a blue plate with white lettering, installed as late as the 1960s — and Ogden's. The Ogden's panels were devised to imitate wood-graining for oak, walnut and mahogany finishes and bore not only the names of cigarette and tobacco brands, but also, like those for 'The News of the World', the name of the shop proprietor.

Garnier's motto 'Nothing too large and nothing too small' is illustrated by an anecdote recalled by the staff there. Sometimes a situation would arise when, for instance, 160' long panels in sections were being processed at the same time as 100,000 2" x 2" licence plates. The biggest signs for advertising display had to be laid out on the road to ensure that all the panels matched up and that the sign was spelt correctly!

Over the last decade, Garnier's has absorbed the work and engaged staff of companies which closed down, namely Chromographic of Wolverhampton and James Bruton and Son Ltd of Palmers Green, both of which were of the same vintage as Garniers.

Garnier's is now the largest manufacturer of reproduction signs which are supplied to Dodo Designs (Manufacturers) Ltd to market both in this country and abroad.

Courtesy Dodo Designs.

Page from Garnier's catalogue, early 1900's.

ENAMELLED IRON PLATES
SINGLE-SIDED. DOUBLE-SIDED. FLAT. BENT. SHAPED
IN ANY SIZE —— IN ANY QUANTITY

HOT WATER NIGHT OR DAY WITHOUT THE USE OF COAL
All Particulars From THE GAS LIGHT & COKE Co. Horseferry Rd. WESTMINSTER S.W.
Actual Size 2 ft. × 1¼ ft.

JOHNNIE WALKER
Actual Size 9 in. × 11 in.

GARNIER & Co Ltd ENAMELLED PLATES — NOTHING TOO LARGE! NOTHING TOO SMALL!! FOR ANY & EVERY PURPOSE — MANUFACTURERS

BUDDLES UNDERTAKER FRONT STREET, NEWBIGGIN. BOARD KEPT HERE
Actual Size 2½ ft. × 2 ft.

Schweppes SODA WATER
Actual Size 2 ft. × 2 ft.

MEMBER OF THE LOCAL LICENSED VICTUALLERS AND BEER RETAILERS ASSOCIATION
Actual Size 13 in. × 11 in.

THE RISING SUN
For Roadside Inns.

MOTORISTS BEWARE OF TRAPS & SPEED LIMITS THROUGH REIGATE — 28 MILES TO HATCHETTS PICCADILLY. W. ACKNOWLEDGED TO BE THE BEST & MOST SELECT MODERATE PRICED RESTAURANT & GRILL ROOM IN LONDON — A la carte - a speciality Table d'hôte Lunch 2/6. Dinner 3/6. Supper 2/6. MUSIC
FIELD PLATES—Actual Size 12 ft. × 8 ft.

Daily Mail LARGEST CIRCULATION
Actual Size 4 ft. 6 in. × 2 ft. 10 in.

PLATES SUITABLE FOR ANY TRADE.

CHEESEMAN & SON FURNITURE REMOVERS Whitton HOUNSLOW — CHEESEMAN & SON REMOVERS & HAULAGE CONTRACTORS
Actual Size 7 ft. 6 in. × 3 ft.
ANY DESIGN, HOWEVER ELABORATE, CAN BE INCLUDED.

Cailler's GENUINE SWISS MILK CHOCOLATE 1D PACKET
5½ in. × 13¼ in.

E. BECKETT & Co. — ESTD 1876
Shaped Plate.

WALTER MOORE & Co.
Bent Plate.

OUR COMPLETE CATALOGUE POST FREE ON RECEIPT OF TRADE CARD.

Page from Garnier's catalogue

Reduced Illustrations of a few cheap and effective forms of Advertising as supplied to the Leading Advertisers.

ADVERTISING FINGER PLATES
ENAMELLED IRON.
One Quarter size – Actual Size.
2⅞ × 8½ in.

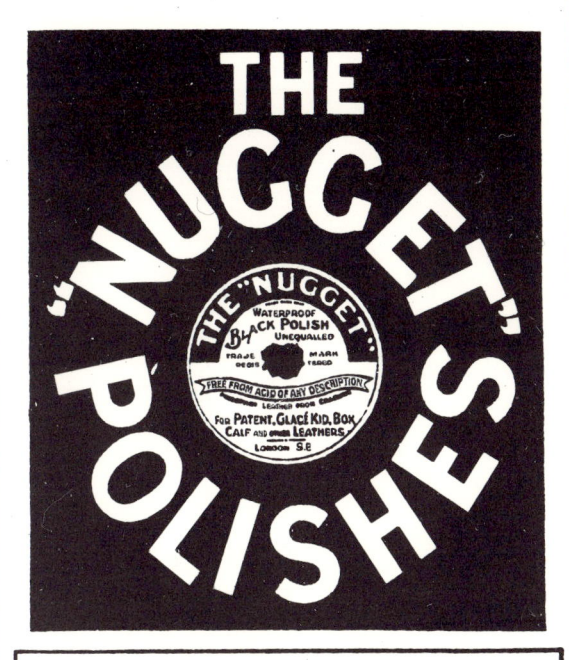

THE BEST BRITISH MADE
LAWN MOWER
"SILENS MESSOR"
GREENS
FOR BRITISH PEOPLE

THE "NUGGET" POLISHES

THE "NUGGET"
WATERPROOF
BLACK POLISH
UNEQUALLED
TRADE MARK
FREE FROM ACID OF ANY DESCRIPTION
FOR PATENT, GLACÉ KID, BOX CALF and other LEATHERS.
LONDON S.E

The above is a Reduced Illustration of a Specimen Set of Enamelled Copper Letters with Enamelled Medallion Fixed on Agents and Shopkeepers Windows throughout the United Kingdom and many places Abroad. Considerably over ONE MILLION of these Letters have been supplied by us to Messrs.
THE NUGGET POLISH Co.
See also Pages 6 and 52.

ADVERTISING MATCH STRIKERS.
ENAMELLED IRON.
(One Quarter size Actual Size.
2⅞ × 8¼ in.)

BEST FOR SONGSTERS
SPRATT'S BIRD SEEDS
STRIKE MATCHES HERE

ENAMELLED IRON ADVERTISING PLATES.

SINGLE-SIDED
MADE IN ANY SIZE OR DESIGN.
One-third size — Actual Size 9 × 3 in.

LLOYD'S NEWS
THE LEADING SUNDAY PAPER

DOUBLE-SIDED WITH FLANGE
MADE IN ANY SIZE UP TO 24 × 18 in.

FAULKNER'S
NOSEGAY
IN PACKETS ONLY

van Houten's
Easily Digested
PURE SOLUBLE
Cocoa
BEST & GOES FARTHEST

OPAL TABLETS
ABSOLUTELY PERMANENT
AS SUPPLIED TO THE LEADING ADVERTISERS.

Table Waters
Schweppes
Gold Medal. PARIS. 1900.

Contracts undertaken for supplying and fixing these and similar plates throughout the United Kingdom at an inclusive charge.

Page from Garnier's catalogue

ENAMELLED IRON ADVERTISING PLATES.

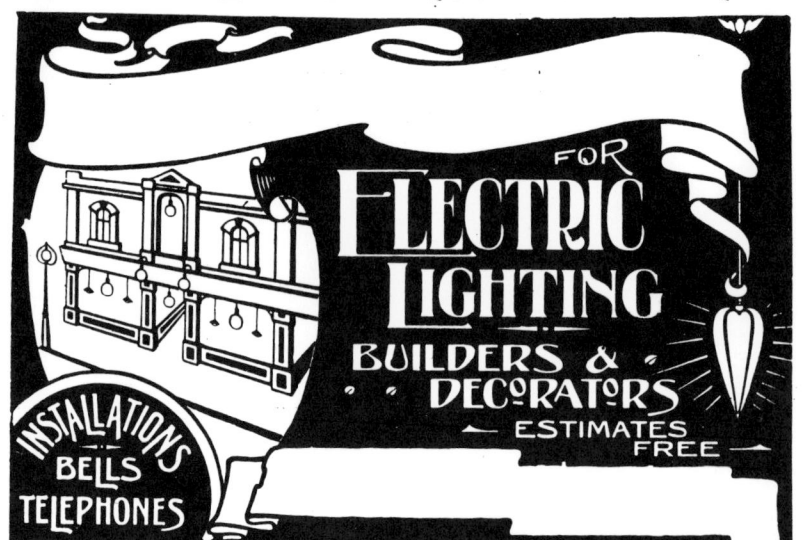

Specially adapted for Electrical Engineers, Builders, Ironmongers, &c., for their own premises or other premises where they have work in hand.

———

Registered Design, E.

(Actual Size, 3' × 2')

UNIQUE DESIGNS AT ORDINARY PRICES.

Registered Design, D.

(Actual Size, 3' × 2')

Any of these Designs may be had in any Two Colours and the blank spaces filled in with Customer's own Name and Address.

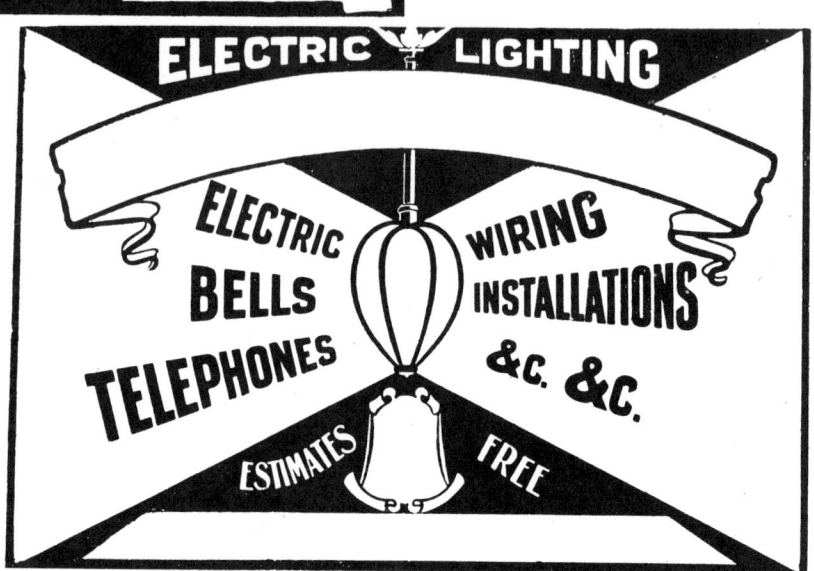

Registered Design, A.

(Actual Size, 3' × 2')

Small Quantities of any of these Designs at a low rate.

———

The Blank Spaces on the Designs are for Customer's Name and Address.

OTHER REGISTERED DESIGNS ON APPLICATION.

74

This cover and the following pages are from the pattern books of the Chromographic Enamel Co. Approximate dates are given, and it can be seen that similar designs were produced over several decades.

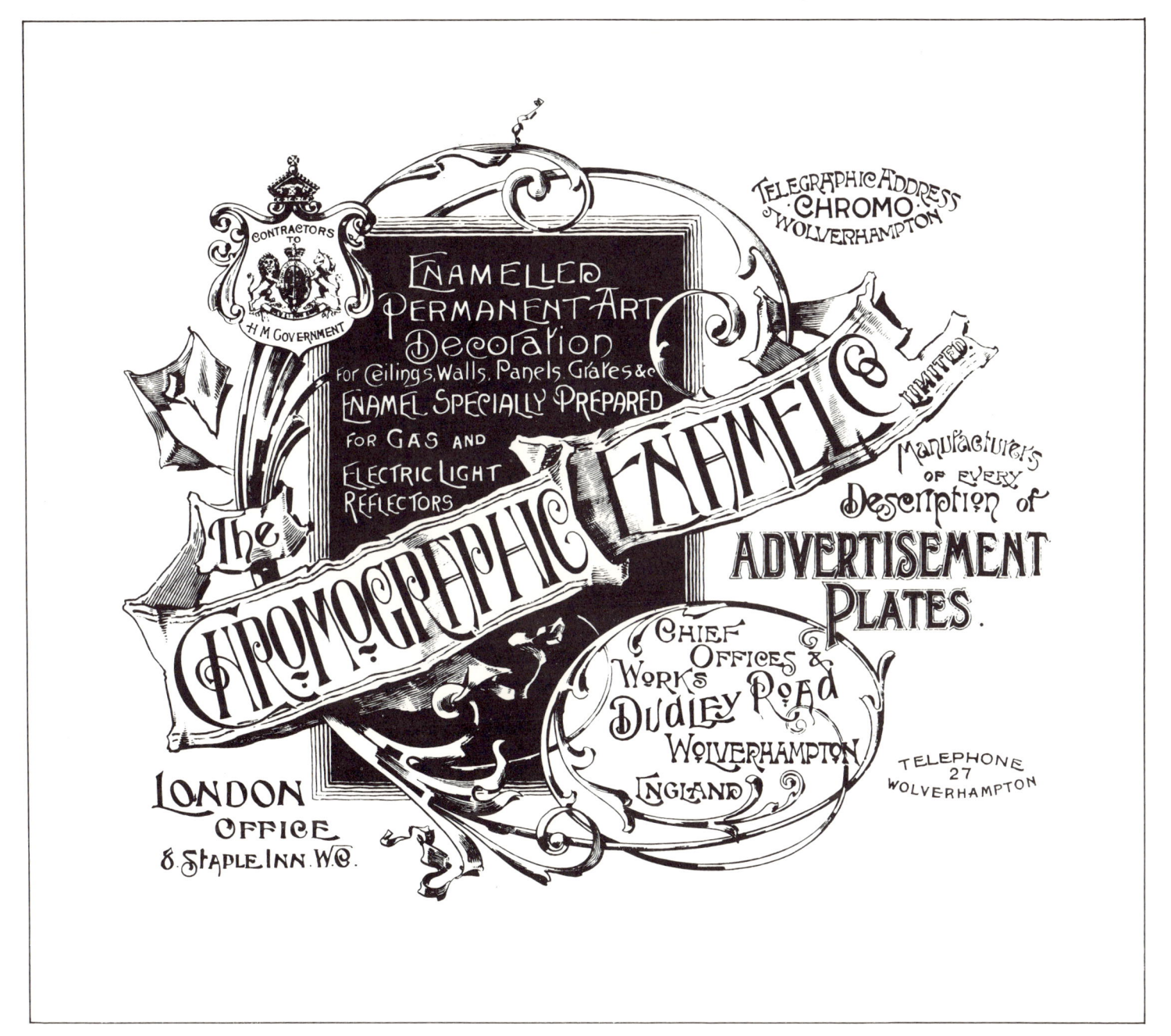

1888.

Circa 1905.

Circa 1911.

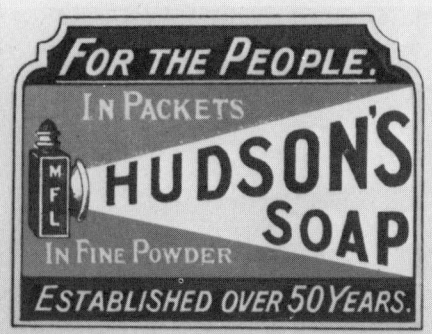

78

Circa 1911.

BEST SOAPS FOR TOILET HOUSEHOLD AND STABLE — **NORTH WEST SOAP COMPY LIMITED CALCUTTA AND MEERUT**
No. 82.—42 × 36 ins.

ZEBRA GRATE POLISH
No. 84 N.—20 × 30 ins.

APPOINTED BY SPECIAL ROYAL WARRANT — SOAP MAKERS TO HER MAJESTY THE QUEEN. **SUNLIGHT SOAP**
No. 83.—36 × 24 ins.

BRASSO METAL POLISH.
No. 85 N.—36 × 8 ins.

Telegraphic Address "FACTORY" — Catalogues Free on application. **DE VILLE & CO. VEHICLE MANUFACTURERS — PAARL**
No. 88.—48 × 36 ins.

Stephen's Inks — 150 140 130 120 110 100 90 80 70 60 50 40 30 20 10 — FEVER HEAT, BLOOD HEAT, SUMMER HEAT, TEMPERATE, FREEZING — For all Temperatures — *Stephen's Inks*
No. 89. (Three sizes.)

BOVRIL
No. 86 N.—48 × 15 ins.

940
No. 93.—9 × 6½ ins.

4352 CONDUCTOR
No. 94.—4½ × 3½ ins.

STAGE 19,565 DRIVER
No. 95.—4½ × 3½ ins.

75
No. 96.—3 × 2 ins.

3
No. 97. 2½ × 1½ ins.

20
No. 98.—6 × 4 ins.

1521
No. 99.—3½ × 1⅞ ins.

10
No. 100.—3⅜ × 2⅜ ins.

30
No. 101.—4 × 2½ ins.

40
No. 102.—5 × 3 ins.

CWTS 21 QRS 3
No. 90.—6 × 3 ins. CART TARE FRAME.

KEEP TO THE RIGHT
No. 91.—5 × 6½ ins. Lettered both sides.

467
No. 92. Spike 3½ × 2½ ins. 9 ins. long.

Circa 1911.

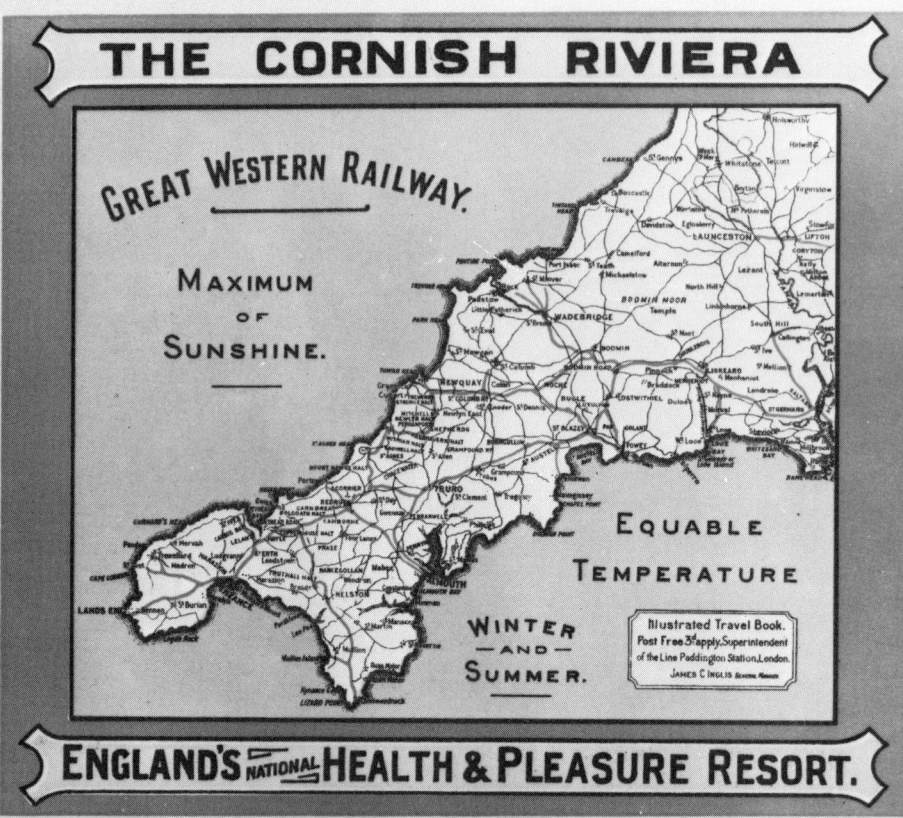

ROBIN STARCH EQUAL TO RECKITTS BLUE No. 1 N.—10×12 ins.	**THE CORNISH RIVIERA** GREAT WESTERN RAILWAY. MAXIMUM OF SUNSHINE. EQUABLE TEMPERATURE WINTER AND SUMMER. ENGLAND'S NATIONAL HEALTH & PLEASURE RESORT. No. 3 N.—14×12 ins. and 42×36 ins.	**GOODALL'S CUSTARD POWDER** MAKES DELICIOUS CUSTARD WITHOUT EGGS IN BOXES 6ᵈ & 1/- EACH GOODALL, BACKHOUSE & Cº LEEDS No. 4.—11×9 ins.
SUTTON'S NOTED SEEDS No. 2.—12×17 ins.		**WEBBS' SEEDS** WORDSLEY, STOURBRIDGE No. 5 N.—24×36 ins.
GUANO OHLENDORFF'S MANURES No. 8.—42×21 ins.	**USE SUNLIGHT SOAP** No. 9.—30 ins. square.	**AGENT FOR P. & P. Campbell Cleaners PERTH** No. 10 N.—20×24 ins. Lettered both sides.	**ROWNTREE'S COCOA & CHOCOLATE CHOCOLATES AND PASTILLES** MAKERS TO H.M. THE KING No. 11 N.—20×30 ins.

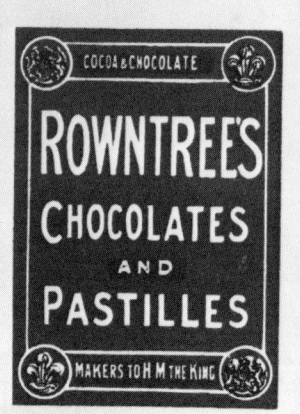

Circa 1911.

KAMATIPURA 12TH STREET
कामाटीपुरा १२ वा रस्ता

F. 17. 24 × 15 ins.

AGENTUR for
Ulykkesforsikrings Selskabet
NOVA.
i HAAG. (HOLLAND)

F. 19. 10 × 8 ins.

شارع الترب
CHAREH EL TOURBAH

F. 18. 24 × 16 ins.

BELGISCH TOL KANTOOR
L'UNION FAIT LA FORCE
DOUANE BELGE

F. 21. 17 × 21½ ins.

पैखाना।
পাইখানা।

F. 20. 20 × 14½ ins.

JALAN UNGKOO ANDOT

F. 22. 19½ × 6¼ ins.

جيد قومان
Colmans Azure Blue
Colman's Blue.

F. 23. 36 × 36 ins.

WINDOW DELIVERY
এইখানে ডাক বিলি হয়

F. 24. 22 × 7½ ins.

BRODIES ROAD
பிரடீஸ் சாலை

F. 25. 30 × 9 ins.

81

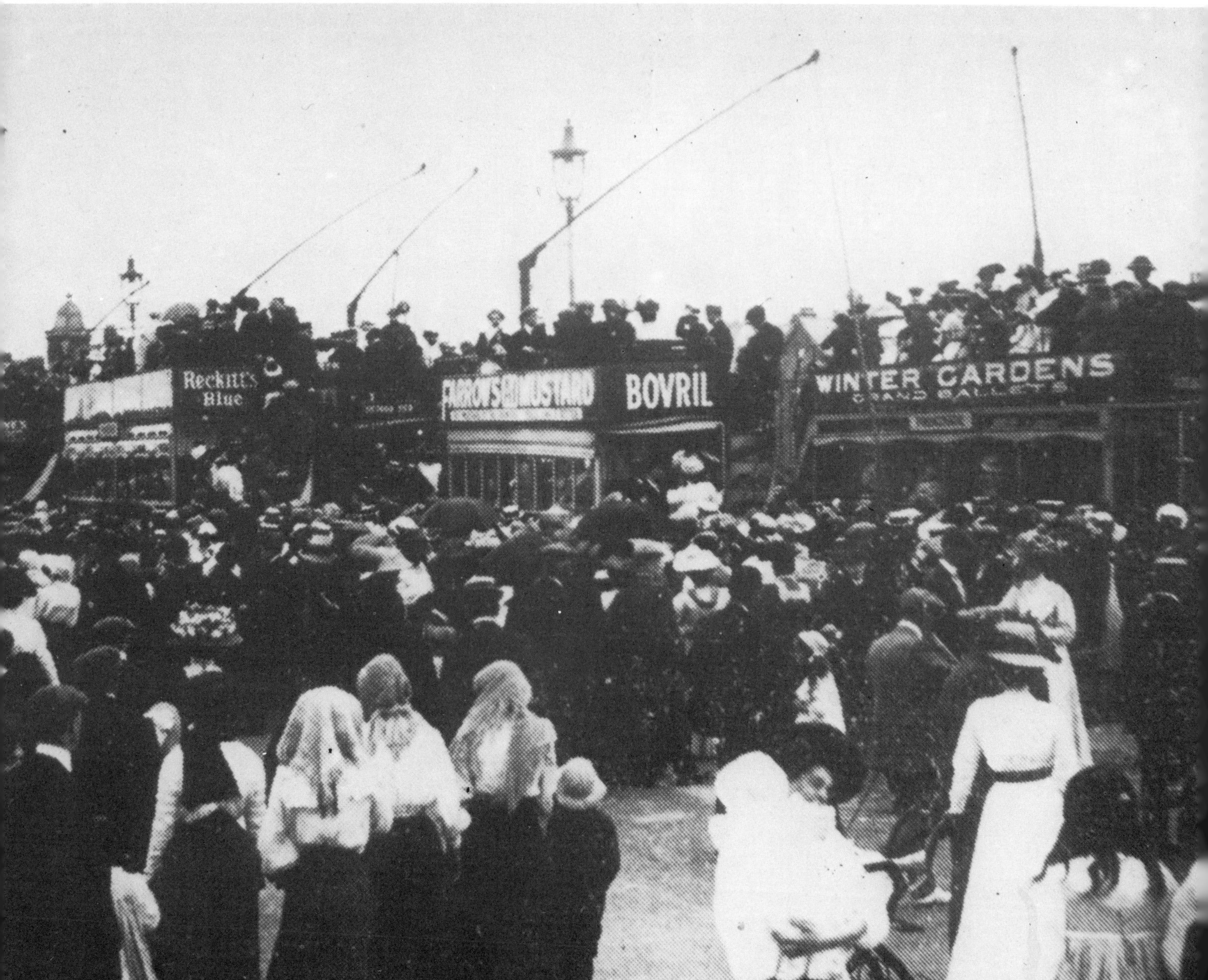
Reckitt's Blue
BOVRIL
WINTER GARDENS
GRAND BALLETS

APPENDIX I
MANUFACTUERS OF ENAMEL ADVERTISING SIGNS

The following is a list of known major manufactuers, past and present, plus a selection of their clients when possible and dates of signs where known.

Company	Location	Product
Artemail	Brussels, Belgium	Spa, Martini
Bruton	London	Lyons, Tea, Fremlins Lager
*Burnham	London	Tizer, Frys
Chromographic	Wolverhampton	Hudsons Soap, Frys 5 Boys, Milkmaid (c.1890), Stranges Oil, Coopers Remedy (Spanish)
*Enamelled Iron	Oldbury	
Ewwisons	Birmingham	
Falkirk Iron	Falkirk	Mobil, Chairman Tobacco
Franco	London	Pratts Motor Spirit (1923)
*Garnier	London	Typhoo Tea, Cadburys, Van Houten's Nosegay
Griffiths & Browett		Brooke Bonds Tea
Griffiths & Millington	London & Birmingham	Danes Anchor Brewery, Ambrose Kid Gloves
Imperial	Birmingham	
Jordan & Sons	Bilston	Stephens Ink Thermometer (1913), Cooperative Society
*Madras Enamel	Madras, India	
Peco	Birmingham	Volkswagen (1959)
Protector	Eccles	BP Motor Spirit (Union Jack), Filtrate Oils
Patent Enamel	London & Birmingham	Quaker Oats (c.1890) Colmans Mustard & Starch (1929), Fry's Cocoa (c.1900) Whites Ginger Beer, Phipps Ale, Elliman's Embrocation (c.1890)
Stainton & Hulme	Birmingham	Players Navy Cut
Stocal Enamels	Burton & Birmingham	Ediswan Lamps, Shell, Guinness
Wildman & Maguyer	London	Sturmy Archer, Mobiloil, Triumph Cycles, Lyons Tea, Nestles Milk
Willings	London	
Woodfield	London	
Wood & Penfold	London	Thorleys Pig Food

* Denotes Companies still in existence

APPENDIX II

PUBLISHED MATERIAL

The Penrose Annual 64 1971 Geoffrey Clarke
Ironbridge Quarterly 1,1 January 1972
Building With Steel November 1972 Geoffrey Clarke
Daily Mail 8 November 1976 Richard Lay
Price Guide to Collectable Antiques 1977 James Mackay
Collector's Year Book 1977 Phillip John
In Trust No.4 May 1977 Christopher Baglee & Andrew Morley
BBC TV 'Look North' 5 May 1977
ITV TV 'Northern Life' 19 May 1977
The Journal 24 May 1977 Peter Anthony
Daily Mirror 10 September 1977 George Thaw
BP Oil News Vol.2 No.9 September 1977 W H S Blakey
The Architect Vol.123 No.9 September 1977 Christopher Baglee & Andrew Morley
Shell Times No.11
Tobacco No.1160 September 1977 Mark Stone
Geordie Life October 1977 Christopher Baglee & Andrew Morley
Old Motor Vol.10 No.6 November 1977
Antique Collector January 1978 Christopher Baglee & Andrew Morley
Design for Vitreous Enamelling May 1978 Design Council and VEDC

APPENDIX III

MUSEUMS AND OTHER PUBLIC DISPLAYS OF ENAMEL ADVERTISING

The following are just a few of the Museums and restored private railway lines which have enamel signs on display.

Beamish North of England Open Air Museum, Stanley, Co. Durham
Beck Isle Museum, Pickering, Yorkshire
Black Country Museum, Dudley, West Midlands
Bluebell Railway, Sheffield Park, Uckfield, Sussex
Castle Museum, York, Yorkshire
Colmans Museum, Norwich, Norfolk
Grindon Museum, Grindon, Sunderland, Tyne & Wear
Ironbridge Gorge Museum, Telford, Shropshire
Keighley & Worth Valley Railway, Keighley, Yorkshire
Landmark Centre, Stirling, Stirlingshire
Myreton Motor Museum, Aberlady, East Lothian
National Railway Museum, York, Yorkshire
Peco Museum, Seaton, Devon
Preston Hall Museum, Stockton-on-Tees, Cleveland
Severn Valley Railway, Bristol, Avon.

Also a national chain of clothes boutiques called SNOB have signs on display in most of their shops.

APPENDIX IV

DEALERS IN ORIGINAL ENAMEL SIGNS

The following is a list of the main dealers around the country who buy and sell enamel advertising signs, plus other advertising material.

Dodo, 185 Westbourne Grove, London W.11.
The Family Trading Company, 58a Turners Hill, Cheshunt, Hertfordshire.
Christopher Gordon, 22 Prospect Road, Moseley, Birmingham.
David Griffiths, 69a Portobello Road, London W.11 (Saturdays only).
Pat Pottle, 111 Upper Street, London N.1.
Tins 'N Tiques, 84 Chelsea Antique Market, London S.W.3.
Mike Marsh, Bermondsey Market, London S.E.1. (Fridays only).
John Hawkins, 69 South Road, Herne Bay, Kent.
Stewart Swinney, 20 Stephen Street, Edinburgh.

APPENDIX V

SOME ADVERTISER'S CLAIMS AND SLOGANS FROM ENAMEL SIGNS

Sisson's Paints make Healthy Homes
Watson's Matchless Cleanser is the best soap for all purposes
Use Watson's Matchless Cleanser and take things easy
Nubolic Disinfectant Soap ensures Sweetness, promotes Health
Lucas Batteries on extended credit terms
Nestles Milk — only on quality, richest in cream
Eat Colman's Mustard with prime beef, mutton etc., and enjoy them more
Mobiloil — make the chart your guide
Drink Tizer the Appetizer
For your throat's sake smoke Craven 'A', they never vary
Hudson's Soap — for the people
Hudson's Soap — powerful, easy and safe
Hudson's Soap — (dry soap) in fine powder, used in all the 'Happy Homes of England'
Player's Please, its the tobacco that counts
Fry's Chocolate. 300 gold medals, by Royal Appointment
Pelaw Liquid Metal Polish, for all metals, brilliant lasting economical easy 2d 3d 6d
Shell shortens every road
Cadbury's Chocolate — delicious and wholesome
Cadbury's Cocoa — absolutely pure, goes furthest
Fry's pure concentrated cocoa
Walter Willson's Smiling service shops for better food
Cogschultz ammunition sold here. British made
Insist upon having Colman's Starch sold in cardboard boxes
Virol — delicate children need it
Oxo — splendid with milk for children
Oxo — excellent with milk for growing children
Guinness is good for you — gives you strength
Daily Mirror — best all along the line
It pays to pay less and fit Lissen mica plugs, guaranteed for one year

We sell sealed Shell
Colmans D.S.F. Mustard
Your chemist sells Numol, the body bone building food
Best in the world Duckham's Adcoids. Save — power pounds patience
Feed your dog on Spillers Shapes 4 shapes 4 flavours
Nectar Tea the most economical tea sold
A packet for every pocket — Lyon's Tea — always the best
Ogden's St Bruno the tobacco that won't be hurried
Desperation, pacification, acclamation, realization. 'It's Fry's' (5 boys)
We close today at...... ¼lb packets Hudson's Soap
Veno's lightening cough cure. Cures coughs, colds and chest diseases. Price 9½d, 1/1½d
Ask us for Rowntree's Elect Cocoa it has the Rowntree flavour
Golden Shred — the world's best marmalade
Ingersoll watches 5/- and up for sale here
Rinso — soak the clothes — that's all! Saves coal every wash day

85

INDEX

Index of text references to signs, advertising companies, sign manufacturers and individuals together with an index of signs illustrated in single colour.

A1 Sauce 28
Aeroshell 50
Andrews Liver Salt 67
Anti-Laria 30
Aquascutum 13
Armco 15
Avon Tyres 56
BP 50, 51, 61
BSA 51
Baugh, Benjamin 12, 70
Beecham's 9, 29
Bensdorp's Cocoa 28
Bermaline Bread 56
Berry's Boot Polish 30
Black Cat 62
Blue Cross 32
Boar's Head Tobacco 71
Borwick's Baking Powder 70
Bovril 82
Brasso 31
Brighton Pier 60
Brilliant Sign Co. Ltd., The 25
Brooke Bond Tea 21, 30, 32
Bruton & Son Ltd., James 12, 71
Bryant & May's Matches 27, 30
Bull, John Tyres 51
Burnham's (Onyx) Ltd. 13, 71
Cadbury's 12, 13, 31, 71
Campbell, Malcolm 60
Camp (coffee) 32, 54, 55
Capstan Navy Cut 32, 62
Castrol 60
Champion Norfolk Boots 57
Chiver's Carpet Soap 31, 66
Chromographic of Wolverhampton 13, 70, 71
Chromographic (catalogue illustrations) 75, 76, 77, 78, 79
Churchman's Number 1 Cigarettes 51
Ciment (Fondu) 50
Cleveland Discol 60, 61
Cogshultz Ammunition 67
Colby's, E.A. 68
Colman's Starch 13, 31, 63
Cooper's Sheep Dip 51, 63
Craven "A" 9, 29, 51, 67
Crossley's (engines) 31
Daily Mirror 20
Daily Chronicle 63, 64
Daily Telegraph 64
Dartnell's Ginger Beer 60
Dewar's 25
Dodo Designs 71
Dorne, James (polish) 31
Dragonfly (motor oils) 23
Ediswan 31
Epps's Cocoa 27, 29, 31
Falkirk Iron Co., The 13
Farrow's Mustard 82
Foyle's 62
Fry's Chocolate 5, 13, 17, 22, 55, 66, 71
Fry's Cocoa 13, 29, 31, 51, 71
Garnier, Charles 70
Garnier's 13, 70, 71
Garnier's Catalogues 72, 73, 74
German Military Signs 60
Gilliard & Co. Ltd. 28
Golden Shred 85

86

Gossages 30, 60, 62
Green & Leddicott's 60
Grove, Edward (Tailor) 28
Guinness 60
Hall's Distemper 61
Hay, P & R (dyers & cleaners) 51, 54, 55, 64
His Master's Voice 57
Holzapfels Compositions 57
Horniman's 32
Hovis Bread Flour Co. 13
Hudson's Soap 12, 13, 24, 29, 30, 31, 60, 62, 63, 65, 66
Humber Cars 60
Imperial of Birmingham 13
Ingersoll Watches 9
Iona Whisky 28
Iron Jelloids 20
Jones' Sewing Machines 67
Kaputine 54
Karpol Car Policy 51
Keen's Blue 26
Komo (metal paste) 30, 31
L.V. Pickle, The 28
Lennox's Whisky 64
Lever, W.H. 20
Lipton's Tea 24, 32, 68
Lissen (plugs) 67
Lloyd's News 70, 71
Lucas Batteries 30, 51
Lucas Spades 55
Lyon's Tea 22, 30, 31, 32, 51, 56, 63
Macniven & Cameron's (pens) 66, 67
'Mangling Done Here' 31
Maples 13
Mazawattee Tea 13, 20, 21, 32, 50, 63
Melrose's Tea 32
Milkmaid 27, 57
Mobil 50
Monsters Pop 63
Morris Cars & Service 51, 67
Morris Insurance 66
National (gas engines) 55
National Sign Company 57
Neaves Food 18
Nectar Tea 12, 13, 30, 32, 50, 55
Nestle's Milk 10, 14, 21, 24, 28, 51
New Cross Empire 28
New Hudson Cycles 29, 67
Newcastle Co-op 67
News of the World 28, 71
Nosegay 50, 54, 63, 70
Nu-Texa 61
Nugget Boot Polish 21, 70
Oakey's (knife polish) 28
Ogden's 53, 54, 55, 71
Opie, Robert 20
Ovaltine 61
Ovoline (oil) 61
Ovum 20, 21
Oxo 21, 29, 52
Palethorpes 51, 54, 66, 67
Palmer Cord Tyres 58
Patent Enamel Co., The 9, 12, 70, 71
Pather Iron & Steel Co. Limited 16
Pears 13, 21, 25, 29, 67
Pepsi Cola 51
Petter (oil) 66, 67
Phipps (ale) 61
Piccadilly 50

Pink's Jams 24
Player's 21, 22, 32, 54, 62, 63
Pratt's 23, 31, 60
Price's (soap) 27
Puck Matches 67
Puritan Soap 57
Raleigh 50, 66, 67
Rajah Cigars 65, 67
Ramsden's, Archibald (pianos) 13
Ransome's 20
Reckitt's 21, 31, 82
Redferns 20
Redgate Table Waters 56
Remington 67
Rinso 30, 67
Robin Starch 30, 31
Rowntree's 13, 31, 54, 57, 63, 65
Royal Daylight 31, 51, 57
Salt's Patent Enamel Works 12
Science (floor polish) 61
Shell 23, 50, 51, 54, 61
Silverbrook Tea 32
Simpson's Whisky 30
Singer (sewing machines) 66, 67
Spillers 50, 57
Spratt's 55
Star 16
Stephens Inks 66, 67
Sturmey-Archer 67
Sun Motor Insurance 60, 67
Sunlight Soap 20, 24, 31, 50, 60, 62, 63, 71
Swan Ink 55, 66
Swan Vestas 57
Taddy's (tobacco) 56
Thomson's Dye 54
Thomson & Porteous 62
Thorley's 13
Tilling, Thos. 32
Tizer 54, 66
Toogood's Seeds 20
Tower Tea 32
Town and Country Planning Act, 1947 21
Turf Cigarettes 21, 22
Turkish Regie Cigarettes 64
Typhoo 32, 71
United Kingdom Tea 32
Van Houten's Cocoa 9, 26, 29, 31, 50, 70
Vauxhall Cars 50
Venus Soap 51
Veritas Gas Mantles 31, 50, 54
Virol 20, 29
Volkswagen 61
Volvolutums 31
Watson's Matchless Cleanser 51
Waverley Cigarettes 56
Waverley Pen 29
Western Laundries 71
Whitbread's 9
Will's 51, 54, 62, 67
Willson's, Walter Stores 51, 70
Wincarnis 29, 30
Winter Gardens 82
Wood & Penfold 13
Woodbine 60, 66
Wright's (soap) 67
Yorkshire Relish 26
Zebo (polish) 31
Zebra Grate Polish 10
Zeppelin Bottles 60

Index of signs illustrated in colour, between pages 32 and 49 listed in order of appearance

Will's Golden Bar
Cadbury's Cocoa
St. Julien Tobacco
Stephens' Ink
Hudson's Soap
Green & Ledicott's
R.D. & J.B. Fraser Ltd.
Colman's Mustard
Karpol
Castrol
Oxo
Will's Woodbines
Matchless Metal Polish

Tizer
Fry's Chocolate
Burgoyne's
Ex-Lax
Hudson's Soap
Royal Daylight
Macniven & Cameron's Pens
Spillers
Rinso
Littlewoods
Thorley's Pig Food
Oxo

Wincarnis
Viking Milk
Player's Navy Cut
National Benzole
Gossages'
Brooke Bond Tea
Sun Insurance
Shell
Hudson's Soap
Lyons Fruit Pies
Fry's Chocolate
Palethorpes

Rubicon Twist
Player's Please
Rowntree's Cocoa
Spratt's
Will's Westward Ho
Blue Band Margarine
Brasso Polish
Will's Star
Will's Woodbines
Thomsons Dye
Rowntree's Chocolates & Pastilles
Veritas Mantles
Dagenite
Palethorpes
Stephens' Inks
Churchman's Tortoiseshell

Chivers' Carpet Soap
Reckitt's Blue
Norfolk Champion Boots
Ovaltine
Kenya Beer
Swan Ink
Morris Service
Patent Steam Carpet Beating Co. Ltd.

Robin Starch
Raleigh Bicycles
United Kingdom Tea
Union Castle Line
Player's Navy Mixture
Mitchells & Butler's Ales
Sharp's Toffee
Ogden's Robin Cigarettes
P & R Hay
Ediswan
Rhodian Curly Cut
Mobiloil
Puck Matches
Bovril
Hudson's Soap
Tizer
Stephens' Inks
St. Bruno
Player's Navy Cut

Spillers Shapes
W.H. Smith
BP
Sunlight Soap
Craven 'A'
Lucas Batteries
Quaker Oats
Silver Shred
Bermaline Bread
Ogden's St. Bruno
Day & Martins
Lyons Tea
Venos
Fry's Chocolate
Good Year

St. Julien Tobacco
Bibby's Cream
Daily Mirror
Golliberry
Brooke Bond Tea
Burnard & Alger's
Fry's Chocolate
Selo Films
Witness Cutlery
Woodbine Cigarettes
Rajah Cigars
Pioneer Cement
Daily Telegraph
Brooklax
Rowntree's Chocolate
Lyons' Ink
Volvolutum

Oxo
Melox Dog Food
Turf Cigarettes
Britax
Battleaxe Bar
Ogden's Walnut Plug
Bovril
Player's Navy Cut

Shell
Elliman's Embrocation
Churchmans
Swan Vestas
Hudson's Soap
Zebra
Guinness
His Master's Voice
Venus Soap
Spiller's

Duckham's Oils
Komo Metal Paste
Milkmaid
AC Plugs
Guinness
Fry's Cocoa
Goodyear
Player's Drumhead Cigarettes
Brooke Bond Tea
Veno's
Walter Willson's
Phipps
Redbreast Flake
Ovum
Shell

Martini
Nestle's Milk
Carrolls Number 1
Coca Cola
'In Brandstoffen'
Pepsi Cola
Cooper
Belga Cigarettes
Cleveland Petrols
Van Houten's Cocoa
Renault
Good Year Tyres
Mobiloil
Spa
Van Houten's Cocoa

Guinness
Redgate
Rowntree's Pastilles
Morris
Matchless Metal Polish
Jones Sewing Machines
Lissen
Adkin's Tobacco
Ogden's Walnut Plug
Lucas
Colman's Starch
Nectar Tea
Player's Airman
Tizer
The Chairman
Holzapfels

Hall's Distemper
Sunlight Soap
Black Cat

Robsons Ltd. Removals & Storage
Mazawattee Tea
J. Leno & Sons
Dartnell's Ginger Beer
Singer Sewing Machines
Cooper's Sheep Dip
Taylor's Depository
Monsters
Veritas Mantles
Ogden's Coolie Cut Plug
Anti-Laria

Strange's Oil
BP White May & Royal Standard Oils
Turf Cigarettes
Camp
Ingersoll Watches
Ogden's Battle Axe
Burma Sauce
Player's Please
Colman's Mustard
Sunlight Soap
Sturmey Archer
Will's Flag Cigarettes
Coolie Cut Plug
Wills's Gold Flake
Brooke Bond Tea
Craven A
Walnut Plug
Kley Kynoch Cartridges
Lyons Tea
Tizer
Colman's Mustard
Huntley & Palmers
Nosegay
Petter Oil
Swan Soap
Pelaw Metal Polish
Pratt's

Hudson's Soap
Fry's Chocolate
Will's Gold Flake
Churchmans Tortoiseshell
Camp
Cadbury's Chocolate
Mazawattee Tea
Waverley Pen
Swan Vestas

Lyons Ice Cream
Bassett's Mints
Lyons Cakes
Ogden's Robin Cigarettes
Watson's Matchless Cleanser
Lyons Fruit Pies
Barbers Teas
Lyons Cakes
Westminster Virginia Cigarettes
Ogden's Robin Cigarettes

Shell
BP

87

EXHIBITION

An exhibition arranged by Christopher Baglee and Andrew Morley will be touring Great Britain during 1978 and 1979. It contains approximately 150 enamel advertising signs covering the majority of sizes, shapes, ages and types described in this book. Virtually all the signs are illustrated in this work.

The exhibition will be on show at the following locations:-

Newcastle Upon Tyne	Laing Art Gallery	April 1978
Sheffield	Mappin Art Gallery	June 1978
Hull	Ferens Art Gallery	August 1978
Dudley	City Art Gallery	September 1978
Nottingham	Castle Museum & Art Gallery	October 1978
Southampton	Bargate Museum	December 1978
Bristol	City Art Gallery	February 1979
London	Geffreye Museum	March 1979
Edinburgh	Canongate Tolbooth	June 1979
Dundee	Barrack Street Museum	July 1979